HAYWIRED

POINTLESS (yet awesome) PROJECTS FOR THE ELECTRONICALLY INCLINED

MIKE RIGSBY

CHICAGO
REVIEW
PRESS

Library of Congress Cataloging-in-Publication Data
Rigsby, Mike, 1950-
    Haywired : pointless (yet awesome) projects for the electronically
inclined / Mike Rigsby.
        p. cm.
    ISBN 978-1-55652-779-1
1. Electronics—Amateurs' manuals. I. Title.
    TK9965.R54 2008
    621.381—dc22

                                                        2008038929

Cover design: Scott Rattray
Cover photograph: Mike Rigsby
Interior design: Scott Rattray
Photo credits: Mike Rigsby

© 2009 by Mike Rigsby
All rights reserved
Published by Chicago Review Press, Incorporated
814 North Franklin Street
Chicago, Illinois 60610
ISBN: 978-1-55652-779-1
Printed in the United States of America
5 4 3 2 1

To my beautiful wife, Annelle Rigsby, who smiles patiently
when the delivery man surprises her with robot dinosaurs
or thousands of pink plastic eggs.

Acknowledgments

I wish to thank Jerome Pohlen, Renee Zuckerbrot, and Laura Di Giovine for making the publishing process a happy experience. Also, Bri and Kathy Watt, Jim Molnar, G. G. Berta, Fran Shahbazian, and Denise Stanton provided valuable assistance and advice.

Contents

Introduction

Science can be worthwhile, even educational, without being serious. *Haywired* is a resource for the fun side of electronics, a guide to construction of odd creations.

When I was a senior in college, I purchased 20 car batteries (1,100 pounds altogether), a used "armored mill motor" (a 475-pound motor encased in ½-inch-thick steel), and a used Volkswagen, and combined them with my own electronics to make an electric car. The overstressed motorcycle chain drive and the all-too-frequent fires made the car something less than practical. Gasoline cost 29¢ per gallon and Rigsby Electric Corporation never posed a threat to General Motors. But electric cars were the future, I was certain.

Later, I invented the "Watagage." The Watagage was a plant moisture detector made by attaching two long coat hanger legs to a plastic egg body with a small light on top. By turning a dial on the egg's face to the proper setting (described in the instruction manual), the light would pop on when the correct amount of water was added to a plant. At anytime except Easter, plastic eggs are sold in minimum quantities of 2,000. I learned that a product selling for $10 in the store will net the manufacturer—me—$5 (under ideal circumstances). For $5, I had to supply parts, labor, instructions, packaging, sales, and delivery—as well as the cost of design. This was a significant money loser, but I had enough plastic eggs to cover my needs for many years.

Back in the early days of computing, a magazine article promised free electronic parts (microprocessor and memory) for people who would use the parts to build an electronic mouse capable of finding its way through a maze. My mouse had to be programmed with toggle switches, and he would forget everything when batteries were pulled from his memory. I carried my marginal genius to compete in

California and New York. The mouse's competition—there were only 10 or 15 mice in the entire country—were complex robots built by university and research lab teams. Their sophisticated mice found the walls using battery-powered lights and photo detectors. When the TV news team turned on their bright video lights, only my blind mouse—he used door spring "whiskers" to feel the walls—could perform without confusion.

To solve a real problem—raccoons in my trash cans—I designed the Raccoon Buster. Using a noisy programmable tank and a self-focusing sonar camera, I created a device which would detect an approaching critter then "charge" with lights flashing and siren blaring. Although Buster drove the critters away, he tended to disrupt sleep and he required daily replacement of 10 rechargeable batteries.

All of these "not quite failures" were instructive. Learning by doing teaches many lessons.

In this book, you will learn how to make a flashlight and model electric car powered by ultra capacitors rather than batteries. Capacitors can be recharged millions of times and accept a charge quickly. An electric car powered by capacitors could stop at a "filling station" and be fully charged in two or three minutes. Ultra capacitors cost more than batteries and are heavier, but that may change. If you want to build a large car, I include a simple method for calculating capacitor size and run time for your upsized vehicle.

Whether your ideas are futuristic or commercially impractical, you can use the construction hints and project details in this book, along with a little wire and your creativity, to make your own awesome projects.

A Few Thoughts on Safety

Safety is the most important aspect of project construction.

Wear goggles! When you solder, drill holes, cut wood, hammer, cut wire, or snap plastic, wear safety goggles. Solder splashes, drill bits snap, saw blades break, hammers shatter, wire and plastic fly. Find a comfortable set of goggles with clear lenses and use them, always.

When soldering connections, even the cold soldering iron gets very hot at the tip. You can't melt solder without heat, and the heat is enough to make a painful burn. After you solder a connection, don't touch that point—or anything around it—for a couple of minutes.

All of these projects are powered by batteries or low-voltage transformers. Follow the recommendations. Using a 9-volt battery instead of 3 volts can damage parts and may create a fire. Yes, you can start a fire with a 9-volt battery.

Scoring plastic requires the use of a sharp blade. Securely fasten the plastic and the straight edge guide. Do not use your fingers to hold the plastic. Do not put fingers in the path of a blade.

Don't work when you are tired. Don't take shortcuts.

Do work under good lights, away from shadows.

Use common sense and take your time. If you get frustrated or aggravated, stop for a few minutes and do something else.

And remember to have fun!

1

Moving Eyeball Picture

When you walk into a room, the moving eyeballs follow you.

When your friends watch the eyeballs move, they'll think it's creepy. When you show them the wires and springs on the back side, they'll think you're a genius. Have you ever untangled a garden hose or a balled-up string of holiday lights? The methodical work in untangling—patiently taking one small step at a time—is the approach needed to build this.

How does this work? We make two spring-loaded eyeballs. When resting, the eyeballs look straight ahead. An electric motor attached to each eyeball can turn the eye left or right. A motion sensor on the right causes the motor to turn to the right. A sensor on the left causes them to turn left. Stand in the middle of the picture (no active sensor) and the eyes go to the center.

Parts List

8$\frac{1}{2}$" x 11" thin plywood

(2) $\frac{3}{4}$" x 1$\frac{1}{2}$" wood, 8$\frac{1}{2}$" long

(2) $\frac{3}{4}$" x 1$\frac{1}{2}$" wood, 9$\frac{1}{2}$" long

(2) 1" x 1" x $\frac{3}{4}$" wooden block

(2) PIR sensor modules, www.parallax.com, #555-28027

(2) Stainless steel extension springs, 0.25 O.D., 2.5" free length, www.smallparts.com, #ESX-0015-02

(2) Relay, 5 VDC DPDT, www.jameco.com, #139977

(1) Slide switch, DPDT, Radio Shack, #275-403A

(2) NPN switching transistors, Radio Shack, #276-1617

(3) Silicon diodes, 200V 1 amp, Radio Shack, #276-1102

(2) 1K-ohm resistors, $\frac{1}{4}$ watt, Radio Shack, #271-1321

(1) 12" red hookup wire, stranded, 22 gauge, Radio Shack, #278-1218

(1) 12" black hookup wire, stranded, 22 gauge, Radio Shack, #278-1218

(4) Screw eyes, size 216 $\frac{1}{2}$" (small), hardware store

Cedar balls, hardware store

$\frac{3}{8}$" wiggle eyes, craft store

Perfboard, 2" x 4", Radio Shack, #276-1395

(2) One-cell AA battery holders, Radio Shack, #270-401

(1) Two-cell AA battery holder, Radio Shack, #270-408

(2) 1.5 to 3V DC metal gearmotors, Radio Shack, #273-258

(4) AA batteries

Solder

Glue

Wood screws

Pushpins

Electrical tape

Braided picture hanging wire

Superglue

Double-sided tape

Tools List

Wood saw

Wire wrap tool (see chapter 5 for tips on wire wrapping)

Soldering iron

Drill

$\frac{1}{8}$", $\frac{3}{16}$", 1", and $\frac{7}{8}$" drill bits

Screwdriver

Pencil

To get started, you'll need to find or print an 8½ × 11 inch photo or drawing whose eyeballs are at least 2 inches apart, measuring from the center of the left eyeball to the center of the right eyeball. You need at least that much room for the motors and springs.

Next, using the templates on page 39, cut and assemble the five wooden pieces as shown below. Drill the 1-inch holes in the side pieces before attaching them to the plywood. Fasten the pieces together using wood screws. This creates the picture frame.

Back

Front

Place the photo on top of the frame, and poke a pushpin through the center of each eye.

Using the pinholes to identify the center points of the eyeholes, drill two ⅞-inch-diameter holes.

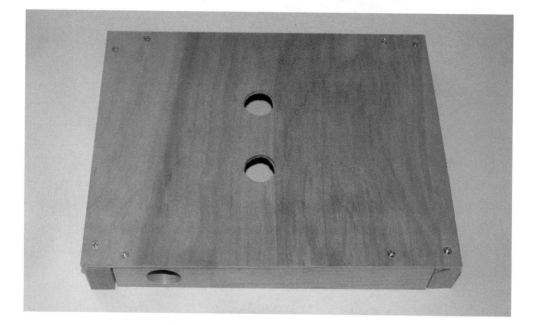

Now, lay the picture facedown and place the frame on top of the picture. Be sure that the bottom of the picture is properly lined up with the bottom of the frame. (If you accidentally match the top of the frame with the bottom of the picture, then the eyeholes in the picture will not match the eyeholes in the frame.) Draw an outline of the circles onto the back of the picture using a pencil.

Cut the eyeholes out of the picture, then place the picture on top of the frame to be certain that the eyeholes match. Once everything matches, place the picture aside until you have finished building the mechanics of this device.

Attach <u>hookup wires</u> (22 gauge) to the motors with solder. (If you are unfamiliar with soldering, see chapter 7.) First, solder a 3-inch-long red wire to the right terminal of each motor. Then solder a 3-inch-long black wire to the left side of each motor. The motors don't care about wire color, but the colors will help you keep track of which wire is which.

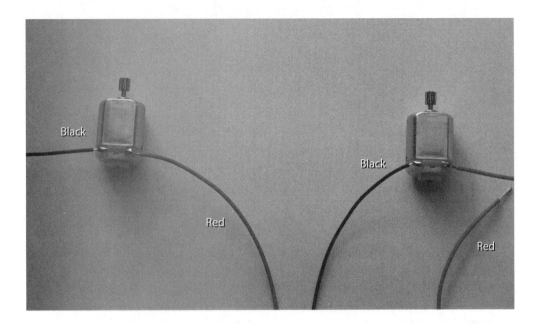

Use round wood balls for the eyes. Cedar balls, "for use in storage chests, boxes, garment bags, etc." are available in most hardware and home supply stores. Drill a ³⁄₁₆-inch hole in one end of each ball, then twist a small screw eye into each ball directly opposite the holes you drilled.

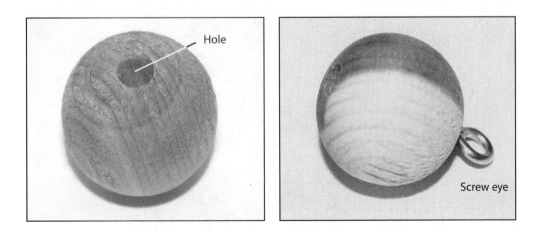

Take two wood blocks (1 × 1 × ¾ inch) and twist a single small screw eye into an end face (a 1 × ¾ inch face) on each block as shown.

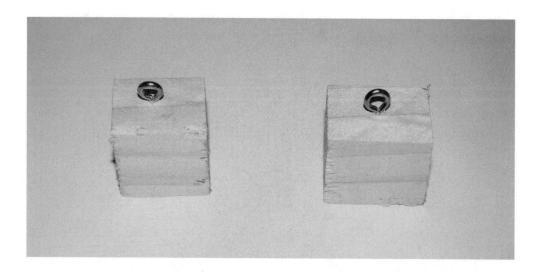

Turn the frame facedown and center a motor beneath one eyehole. Take a pencil and mark the sides of the motor. Repeat with the other eyehole.

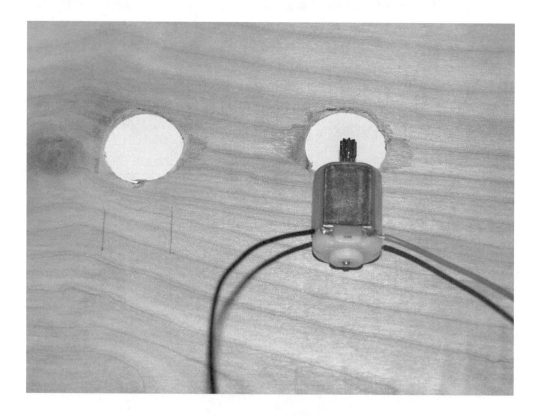

Drill a ⅛-inch hole through each of the four pencil marks.

Push one "eyeball" hole onto the gear of a motor. Attach the spring to the "eyeball" eye screw and—at the other end of the spring—screw the wood block into the frame, screwing through the plywood from the front. The spring should be in slight tension, pulled just beyond the "at rest" position.

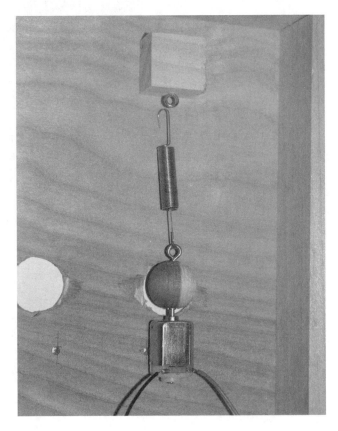

Twist braided picture-hanging wire around the motor to hold it in place.

Solder the springs to the screw eyes. Repeat the process with the other spring, eyeball, wood block, and motor.

Turn the picture faceup and glue wiggle eyes to the front of the eyeballs. The wiggle eyes are necessary since they serve as a "stop" to keep the eye from turning too far or spinning around. Make sure that when the wiggle eyes turn left or right they will make contact with the edge of the holes in the plywood.

Next, glue—superglue works well—three AA battery holders (one two-cell and two one-cells) to the inside wall of the frame. Place the two one-cells side by side.

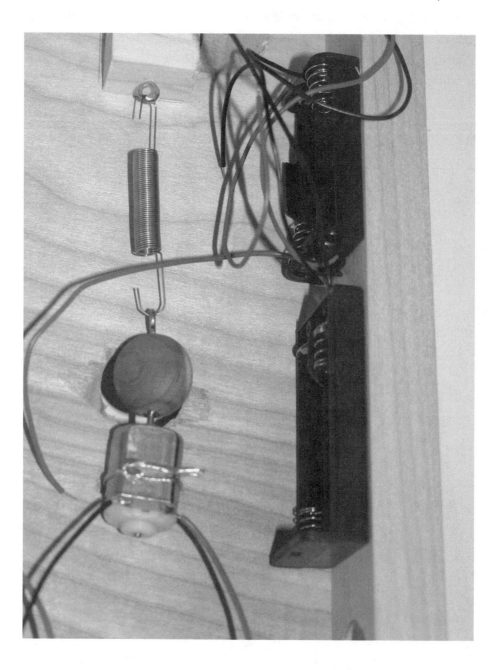

Take a piece of Perfboard and glue the two relays, three diodes, two resistors, and two transistors legs up to the board, as shown. It is not a normal practice to glue all the components with their legs sticking up (like dead bugs). I am doing this so that you can see where the wires are attached. The normal practice (placing the component legs through the Perfboard), creates a more secure attachment with no glue, but it complicates the matter of figuring out which component leg is which. Be sure that the band on the diodes faces the top of the board. The flat part of the transistor should face the bottom of the board. The coil of the relays (two legs that are separated from the other six legs) faces the bottom of the board.

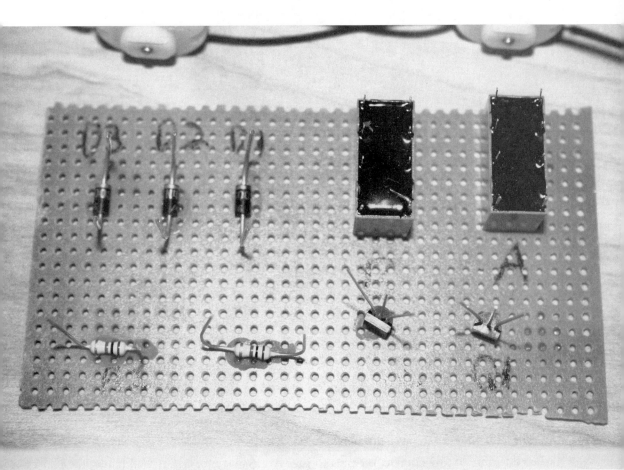

Glue the board to the frame just below the motors.

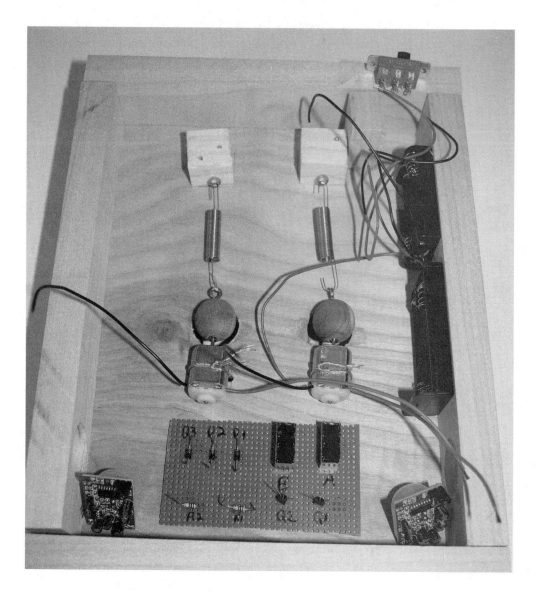

Now it's time to connect the components. This is not difficult, but it is best to take your time. If any connection goes to the wrong place, or is not well made, the project will fail.

Refer to the schematic diagram on the following page to understand where the connections are made. I have identified each connection with a number. In the photographs, the end of the soldering tool (or the wire wrap tool) will be resting on the connection point.

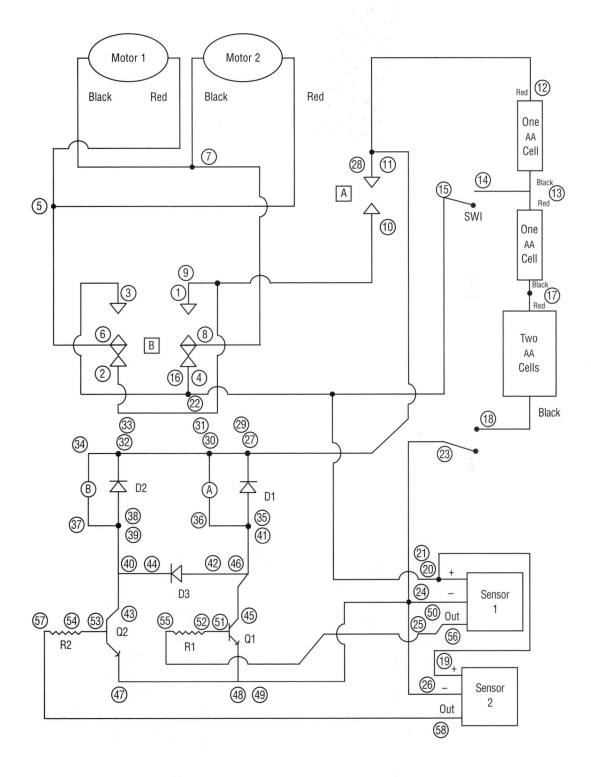

Refer to the following pin diagram for the switch, relays, and transistors.

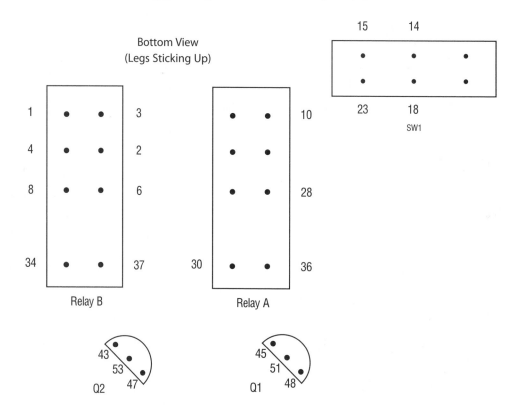

Connection #1 goes to Pin #2 on Relay B.

The other end of this wire, Connection #2, goes to Pin #1 on Relay B.

Connection #3 (one end of a new wire) goes to Pin #4 on Relay B.

The other end of this wire, Connection #4, goes to Pin #3 on Relay B.

Connection #5 connects the two red wires (one from each motor) with a wire-wrap wire. Solder this connection together.

Wrap this connection in electrical tape so that it will not accidentally come in contact with other connections.

The other end of the wire-wrap wire, Connection #6, should be connected to Pin #8 on Relay B.

Connect the two black motor wires to a wire-wrap wire with solder to make Connection #7. Tape the connection with electrical tape.

Attach the other end of the wire-wrap wire, Connection #8, to Pin #6 on Relay B.

Connect one end of a wire-wrap wire to Pin #2 on Relay B (photo on page 16). This is Connection #9. Connect the other end of this wire to Connection #10, Pin #10 on Relay A.

Connection #11 goes to Pin #28 on Relay A.

Take the other end of the wire-wrap wire and solder it to the red wire from the one-cell AA battery holder to form Connection #12. Wrap this connection in electrical tape after it is soldered.

Take the black wire from the one-cell AA battery holder (from the previous step) and connect it to the red wire from the other one-cell AA holder. Also, connect one end of a wire-wrap wire at this point. Solder them all together, then tape the connection. This is Connection #13.

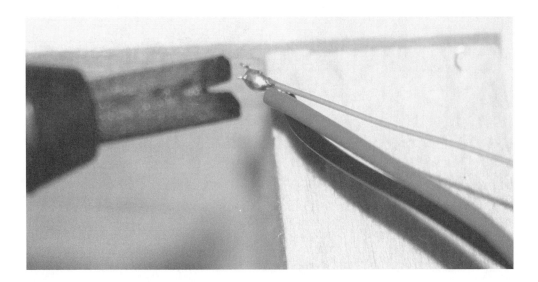

Solder the other end of the wire-wrap wire to a center terminal on switch 1 (see diagram on page 16) to form Connection #14.

For Connection #15, solder one end of a wire-wrap wire to Pin #15.

Connection #16 goes to the same place as Connection #4 (Pin #3 on Relay B). For Connection #17, take the black wire from the other one-cell AA holder and solder it to the red wire from the two-cell AA holder. Tape this connection.

For Connection #18, take the black wire from the two-cell AA holder and solder it to the switch.

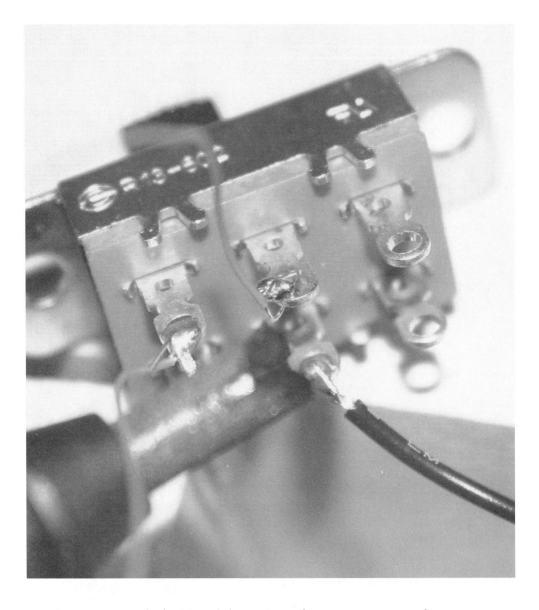

Connect one end of a 10-inch-long piece of wire-wrap wire to the center post of Sensor Module 2 for Connection #19. Connect the other end of this wire to the center post of Sensor Module 1. This will be Connection #20.

Attach one end of a piece of wire-wrap wire to the same place as Connection #20 to make Connection #21.

Connection #22 (the other end of the wire for Connection #21) attaches to the same place as Connection #4 on Relay B (see photo on page 18).

Connect one end of a piece of wire-wrap wire to the switch at Pin #23 to form Connection #23.

The other end of this wire, forming Connection #24, goes to the negative post of Sensor Module 1.

Connection #25 goes to the same place as Connection #24 above. Connection #26 goes to Sensor Module 2, in the same location as in the photo above.

For Connection #27, attach one end of a wire to the cathode (the end with a stripe) of Diode D1 using the wire wrap tool.

Attach the other end of this wire to the same place as Connection #11 (see page 21).

Take one end of a new piece of wire and attach it to the same place as Connection #27 (see page 26) to form Connection #29. Connection #30 goes to Pin #36 on Relay A.

For Connection #31, attach one end of a new piece of wire to the same place as Connection #30.

Connection #32 attaches to the cathode (striped end) of Diode D2.

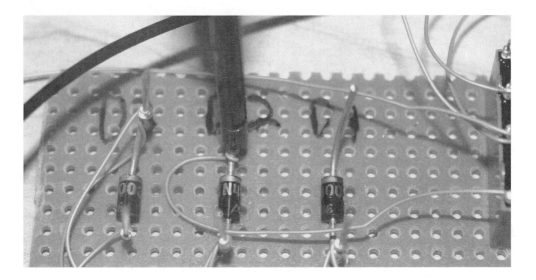

Connection #33 goes to the same place as Connection #32 (see page 27). Connection #34 goes to Pin #34 of Relay B.

Connection #35 connects to the anode (non-band end) of Diode D1.

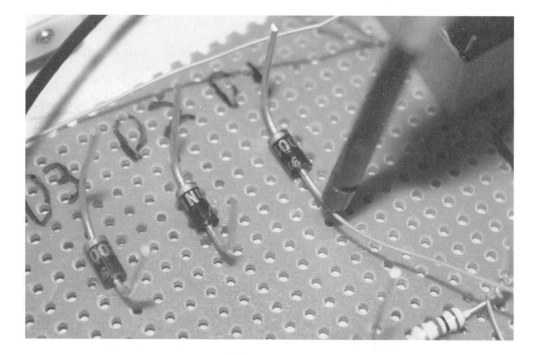

Connection #36, the other end of Connection #35 wire, goes to Pin #36 on Relay A.

Connection #37, one end of a piece of wire, goes to Pin #37 of Relay B.

Connection #38, the other end of this piece of wire, goes to the anode (nonstripe end) of Diode D2.

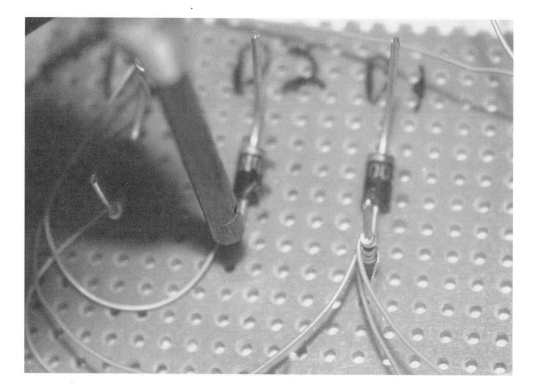

Connection #39 is attached to the same place as Connection #38 in the photo above. The other end of this wire, Connection #40, goes to the cathode (stripe end) of Diode D3.

Connection #41 goes to the same place as Connection #35 (see page 28). Connection #42, the other end of Connection #41, goes to the anode (nonstripe end) of Diode D3.

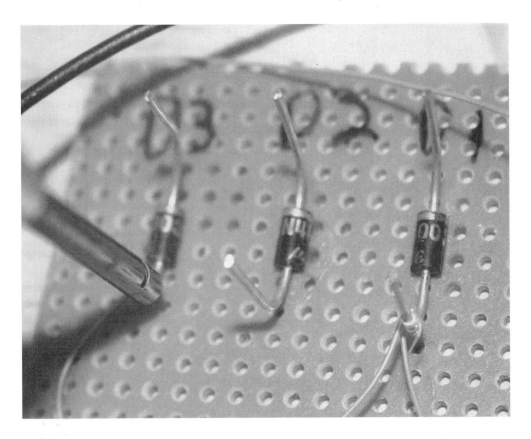

Connection #43 starts at Transistor Q2, Pin #43.

Connection #44, the other end of Connection #43, goes to the same place as Connection #40, page 30.

Connection #45 starts at Transistor Q1, Pin #45.

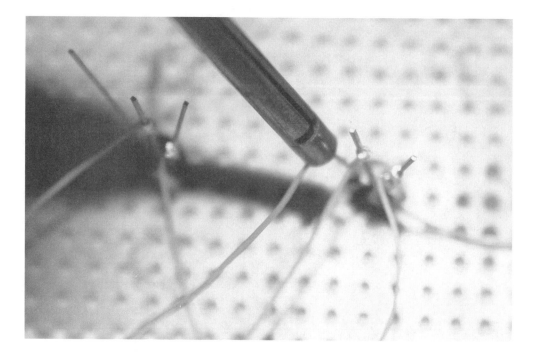

Connection #46, the other end of the wire at Connection #45, goes to the same place as Connection #42, page 31.

Connection #47 starts at Transistor Q2, Pin #47.

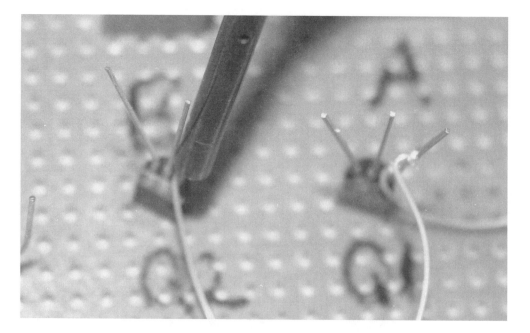

The other end of this wire, Connection #48, attaches to Transistor Q1, Pin #48.

Connection #49 attaches to the same place as Connection #48 above.

Connection #50, the other end of the wire starting at Connection #49, attaches to the same place as Connection #24, page 26.

Connection #51, a new piece of wire, connects to the base of Transistor Q1, Pin #51.

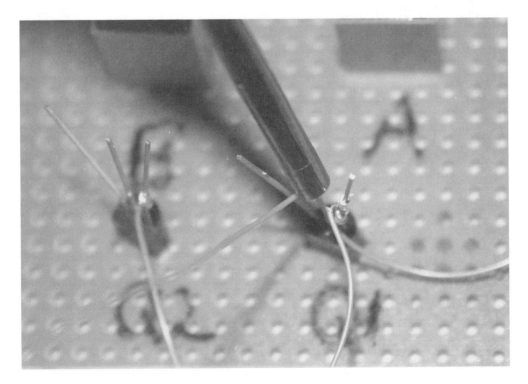

The other end of this wire, Connection #52, attaches to one end of Resistor R1.

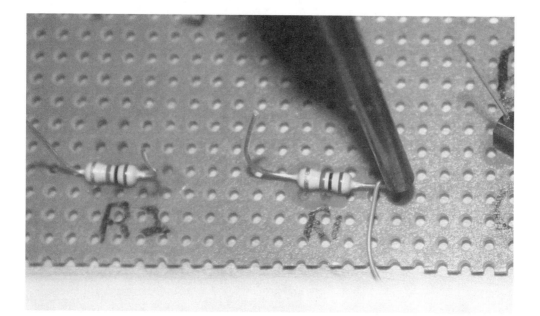

Connection #53 begins at the base of Transistor Q2, Pin 53.

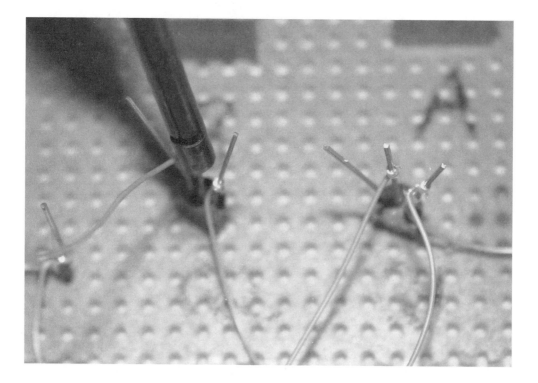

The other end of this piece of wire, Connection #54, attaches to one end of Resistor R2.

Connection #55 starts the end of Resistor R1.

The other end of this wire, Connection #56, attaches to the output of Sensor Module 1. This sensor module is placed on the right-hand side of the frame (as viewed from the back).

Connection #57 attaches to the end of Resistor R2.

Connection #58, the other end of the wire, attaches to the output of Sensor Module 2, as in the photo on page 36. This module is placed on the left-hand side of the frame (as viewed from the back).

Now you are ready for a test of your circuit. Put four AA batteries in the holders and turn the switch on. The switch is on when the black slider is pushed in the direction of all the soldered switch connections. The switch is off when the black slider is pushed in the direction of the two unused pins.

Allow the sensors to "adapt" for about 30 seconds. Place your hand in front of one of the sensors. Hopefully, both eyeballs will move together in the direction of the sensor. If the eyes move the wrong way, it's not too hard to fix. Remove Connection #56 and Connection #58. Place Connection #56 on the pin that used to hold Connection #58, and place Connection #58 on the pin that used to hold Connection #56.

Place a drop of superglue on the white part of the right-hand sensor. Insert the right-hand sensor into the hole on the right side of the frame. The sensor should bond to the frame. Repeat with the left sensor. Add two screw eyes and braided picture-hanging wire. Use double-sided tape to attach the picture to the front.

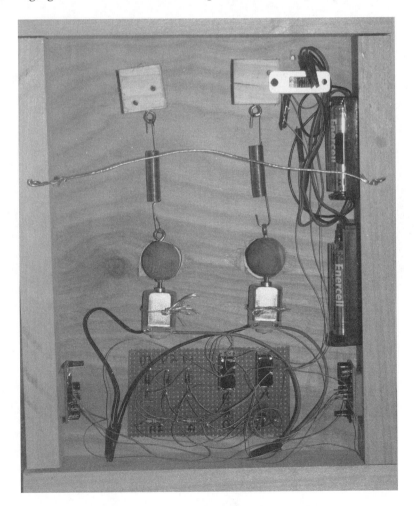

You've done it. Now it's time to show your friends!

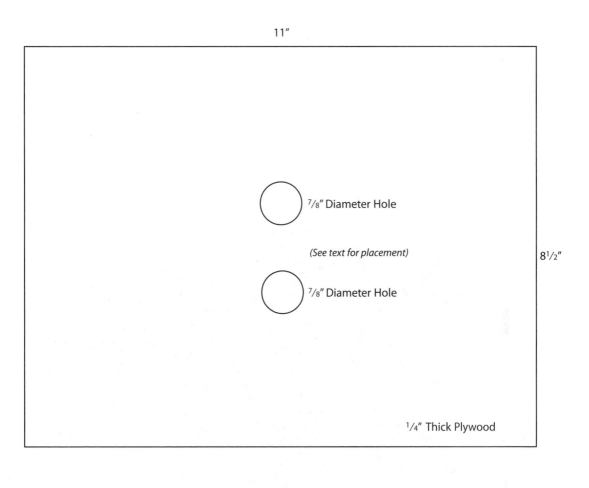

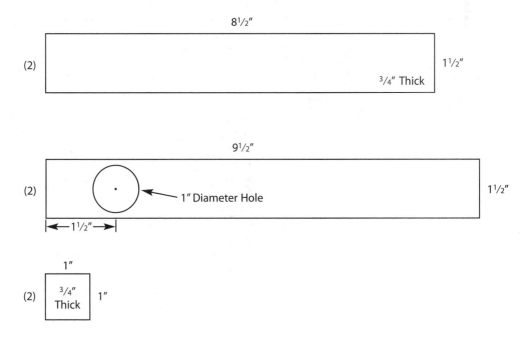

Details

Each sensor module has a setting that you may need to adjust. In the corner, there are three pins, partly covered by a small "jumper." You can pull the jumper straight off with your fingers. The photo below shows the jumper removed.

Place the jumper across the H and middle pin. In this position, the output stays high for a few seconds when someone enters the detection area.

Why, you ask, are there so many parts? To make the motors turn one way or the other, you have to be able to reverse the positive and negative wires to the motor. This is done with the relays. Although the relays don't require much power, the sensors, which you need to detect motion, are major wimps when it comes to having enough power to do anything. You have to take the tiny output signal from the sensor and make it operate a relay. The output from the sensor goes through a resistor to a transistor. The resistor protects the wimpy sensor from burning itself out. The transistor turns on the relays, which operate the motors.

Why does the circuit have Diodes D1 and D2? The relays don't turn off easily and sometimes create enough voltage to destroy a transistor. The diodes allow this unwanted voltage to run around in a circle for a few millionths of a second without harming anything.

When Sensor 1 is active, it turns on Transistor Q1, which energizes Relay A, which allows power through the normally closed contacts of Relay B, which causes the motors to turn. When Sensor 2 is active, it turns on Transistor Q2, which energizes Relays A and B. (Diode D3 allows this to turn on both relays; it prevents Q1 from turning on both relays.) When Relay A is energized, it provides power to Relay B. When Relay B is energized, it changes the direction of power flow to the motors, causing them to go the opposite direction from the flow when Relay A alone is energized.

When neither sensor is active, the relays are at rest and no power goes to the motors, so the spring returns the eyes to the center position.

2

Electrical Basics

In simple terms, you need to understand three electrical terms: *voltage* (V), *current* (I, measured in amps or milliamps), and *resistance* (R, measured in ohms).

Voltage can be thought of as the "desire" of electrons to move. Consider two islands in a stormy, shark-infested sea. One island is inhabited by humans but is barren. The other island is covered in coconut trees and other edibles. The humans want to get to the food, but they can't. High voltage is like high desire—nothing happens as long as the sea is filled with sharks. Everybody just sits there, but there is potential for something to happen.

Now, let's say the people on the barren island built a narrow, rickety rope bridge between the islands. They will start crossing, though just a few at a time. The movement across the bridge is current (amps). The narrow bridge (wire) has a high resistance (ohms) because it will allow only a few people to cross at a time. If the desire of the humans to get to the food is greater (higher voltage), more of them will crowd and push their way across the bridge (higher current).

If they build a wide, modern bridge between the islands (R, resistance is low), lots of people will cross at a time (high current).

This takes us to a basic formula: $V = IR$ (Voltage = Current x Resistance). Resistance of things like wire or lightbulbs is usually fixed, so increasing voltage will increase current. The movement of current is what makes things happen in the electrical world. Voltage is just "desire." For example, if you scuff your feet on a carpet on a cold winter's day, then touch a doorknob, sparks will fly. You have created very high voltage, but a very tiny current flows. You get shocked but not injured. The same high voltage exists on the power lines running along the highway. However, touch one of those and lots of current (and damage) will follow.

The cold solder tool described in a later chapter uses a few volts and very low resistance to cause high current flow. The high current flow generates heat to melt solder.

All of the projects in this book are DC (direct current). Batteries supply direct current, meaning that wires connected to the positive (+) terminal will always carry current in one direction only. Most electronic components can only accept current flowing in one direction and will be damaged if the positive (+) and negative (−) connections are reversed.

Just for your information, the power coming out of outlets in your house is AC (alternating current). Each wire changes polarity (+) to (−), 60 times per second. AC power can be moved from the power generation station to your home with very little energy loss along the way. DC power (because it is usually low voltage) cannot be moved long distances without tremendous loss in the wires.

3

Multimeter

A low-cost multimeter is useful when working on electronic projects. (Radio Shack #22-218 is a nice version.) For example, sometimes you lose track of which wire is (+) and which is (–). For these projects, set your multimeter to a low DC volts scale. The low scale on this multimeter is 15 volts.

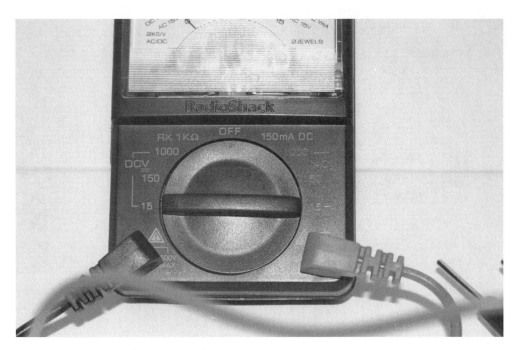

Multimeter set to DC volts, 15-volt scale.

Next, take a 9-volt battery and place the red lead on (+) and the black lead on (–). The meter will display about 9 volts.

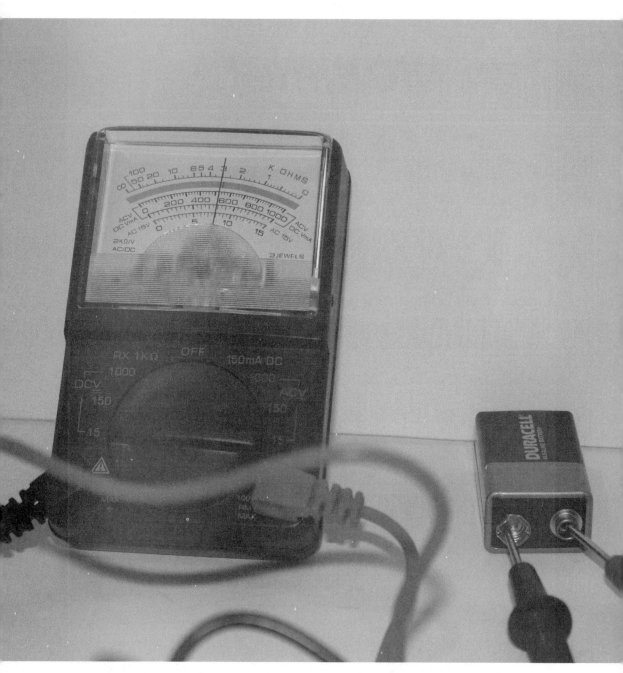

Meter displays about 9 volts.

Switch the leads. Put the red lead on (–) and the black lead on (+). The meter will display less than zero, indicating that the (+) lead is not on the (+) source.

Meter displays less than zero.

To take voltage readings in your circuit, place the black lead on (–) at the battery (or power supply). Place the red lead at the point where you want to measure voltage. If you expect to read (+) volts but nothing happens, then you have to troubleshoot your work.

Multimeter on ohm scale.

A multimeter often is used to check whether connections are good. First, set the multimeter on the lowest available ohm scale. The symbol for ohm looks like an upside-down horseshoe.

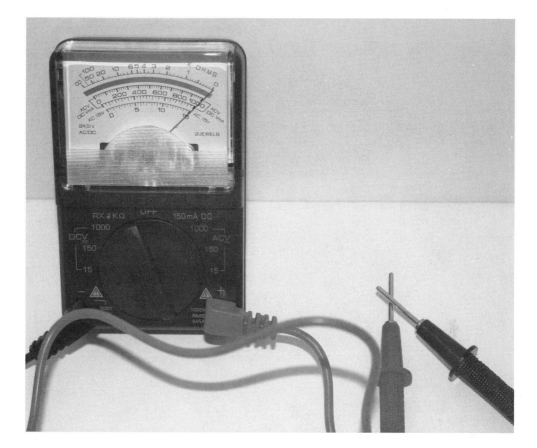

Touch the red and black leads together, and the pointer on the multimeter will move all the way to the right.

Sometimes you will make a poor connection or break a wire. To find bad connections, remove power from the circuit by removing the battery or unplugging the power supply. Put the multimeter on the lowest ohm scale and touch two points that are supposed to be connected. If the pointer moves all the way to the right, the connection you are checking is good. If it is not, find the bad connection and correct it.

4

Talking Room Alarm

Sometimes, people forget to stay out of your room. This project will remind an intruder that he or she has opened the wrong door.

Parts List

9-volt recording module, Radio Shack, #276-1323

9-volt battery

Wire-wrap wire

(1) 1" rectangle ceramic magnet, Radio Shack #64-1879

(2) 3M removable hanger strips

SPST reed relay, www.jameco.com, #111448

Solder

Ruler

Tools List

Soldering iron

Wire wrap tool

First, record a message using the recording module: "Excuse me, but you have mistakenly entered the wrong room. Get out!"

Next, solder one end of a 5-inch-long wire-wrap wire to the place where the black wire from the battery enters the circuit board.

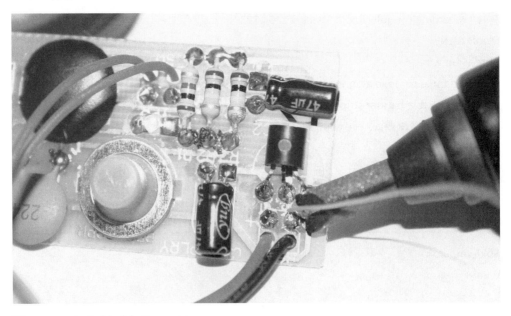

Wire-wrap wire to black battery wire.

Solder the second 5-inch-long wire-wrap wire to the end of R3 as shown.

You now have two wire-wrap wire ends that are not attached to anything. At one end of the reed relay you will find only one pin. Wrap one wire (either one) around this pin. At the other end of the reed relay are three pins. Wrap the unattached wire to the center pin.

Attach a 9-volt battery to the battery socket. Place a magnet near the reed relay. The alarm phrase you recorded should play. The reed relay we used is quite sensitive to magnetic fields, so it works well for this purpose.

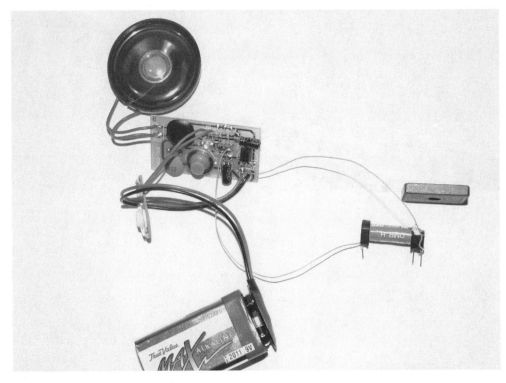

Test the alarm.

Attach two removable hanger strips to the bottom of a door as shown.

These strips can be removed without harming the door if you decide to remove the alarm at a later date. To remove the strips, pull the small stretchy tab until the strips release.

Wrap the battery wires around the upper strip. Insert the relay into the lower strip.

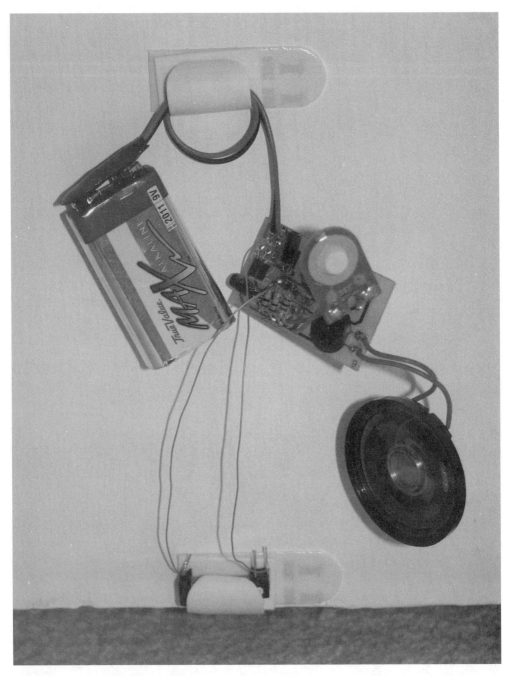

Attach alarm to door.

Place the magnet inside the protected room, in the path of the relay.

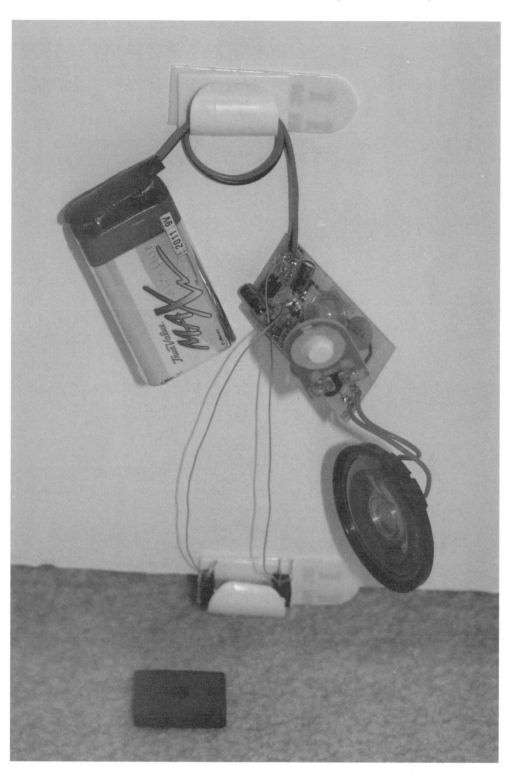

If you are outside the room and you want to enter without setting off the alarm, place a ruler under the door and bump the magnet before opening the door.

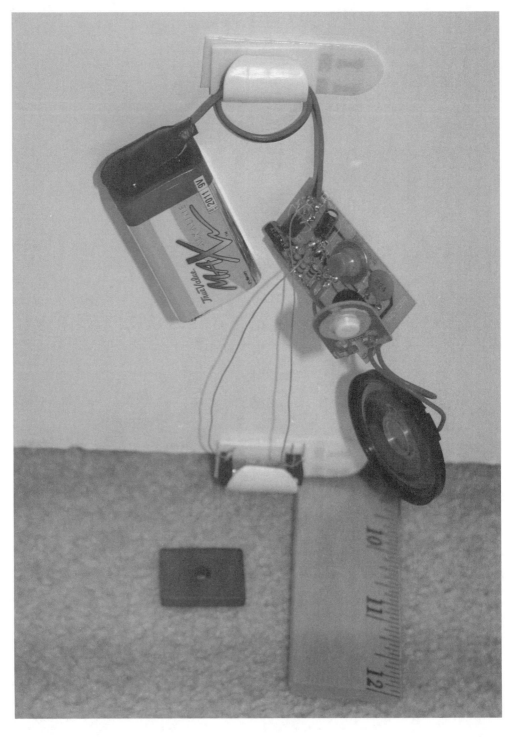

5

Wire Wrapping

Wire wrapping is a technique using special wire and a special tool (Radio Shack #276-1570) to make connections on small electronic devices. The tool has a special shaft that looks something like a screwdriver as well as a wire-stripping tool.

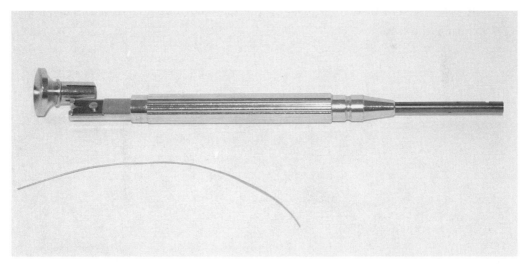

Wire wrap tool and piece of wire

Wire-wrap wire is insulated (electricity can't get through the insulation), so you have to strip off some of the insulation before you can make a connection that will allow electricity to pass. To do this, place about ½ inch of wire through the cutting blade.

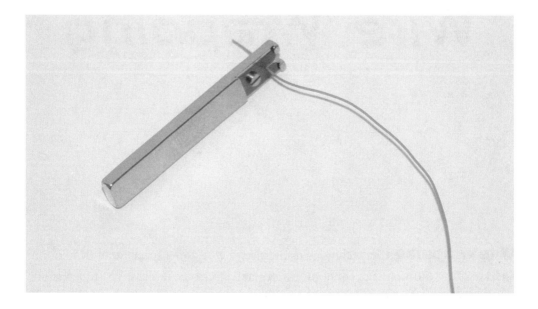

Pull the wire. The insulation (usually a bright color like blue or red) will come off, exposing the wire.

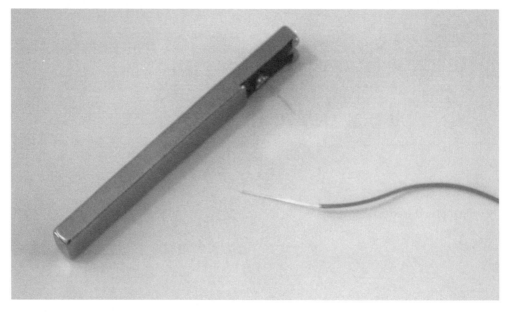

Stripped wire

The end of the wire wrap tool has two holes. The tiny hole on the outer edge is the place you insert the stripped end of the wire-wrap wire. The center hole is placed over the end of the lead on the electronic device you are attaching the wire to. When the electronic device lead is inserted into the center hole, rotate the tool several times and the stripped end of the wire-wrap wire will tightly wrap around the lead.

End of wire wrap tool

Always remember to remove the power source before making any connections with any tool.

6

Talking Greeting Card

In this project, you will construct a talking greeting card. You will be able to record a message up to 30 seconds long.

Parts List

(2) Cardboard pieces, 5" x 7"

Cardboard piece, 5¼" x 7"

(2) Paper pieces 5" x 7"

(1) Paper piece 5¼" x 7"

Glue

9-volt recording module, Radio Shack, #276-1323

(2) CR2032 coin cell batteries

(2) Coin cell battery holders, www.jameco.com, #236996

Lever switch, Radio Shack, #275-016A

Solder

Wire-wrap wire

Tools List

Pen

Scissors

Soldering iron

Wire cutters

Art supplies to decorate card

Start with one 5 × 7 inch piece of cardboard. Place the recording module, switch, and battery holders on the cardboard. Trace an outline of these objects on the cardboard.

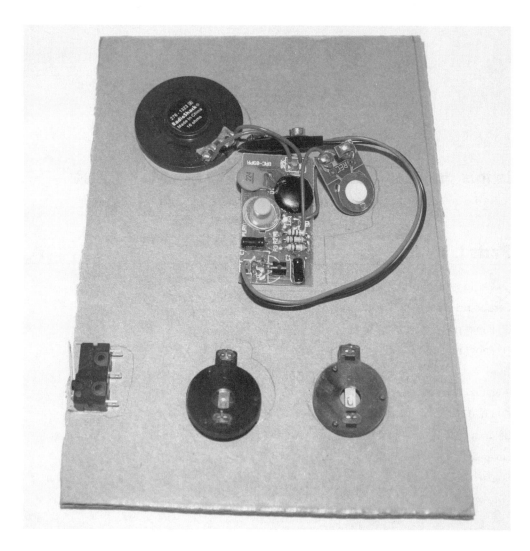

Cut holes in the cardboard around the outlines so that the components have a place to "hide."

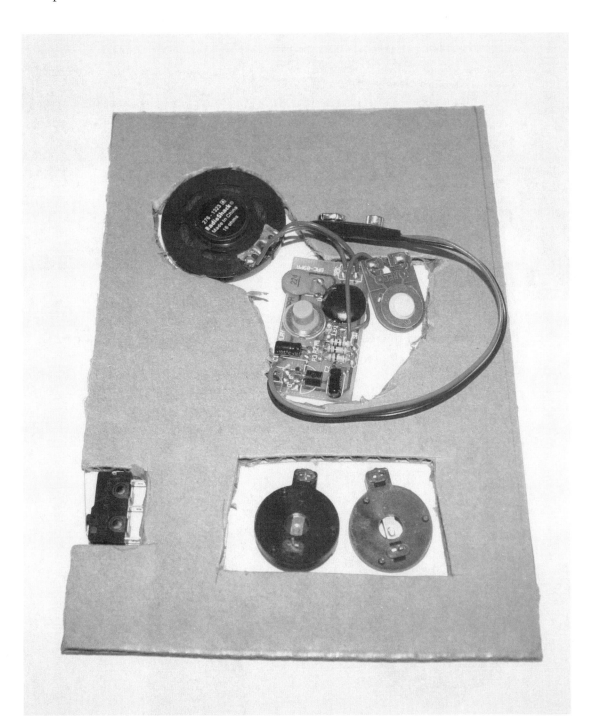

Cut the other 5 × 7 inch piece of cardboard identical to the first one, including the holes. You will use it in a later step.

Glue a 5 × 7 inch piece of paper to the back of the first piece of cardboard.

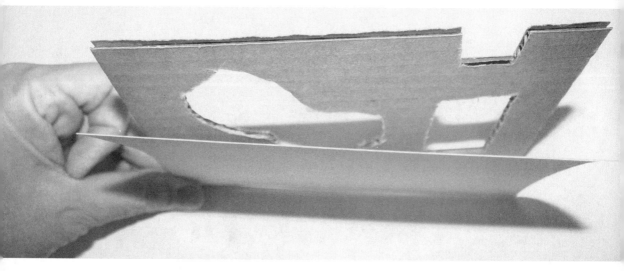

Press the button on the recording module and record your message. Be sure to place the small speaker close to your mouth.

Recording module

On the back of the recording module are two metal tabs. Bend those tabs up and push them through the board.

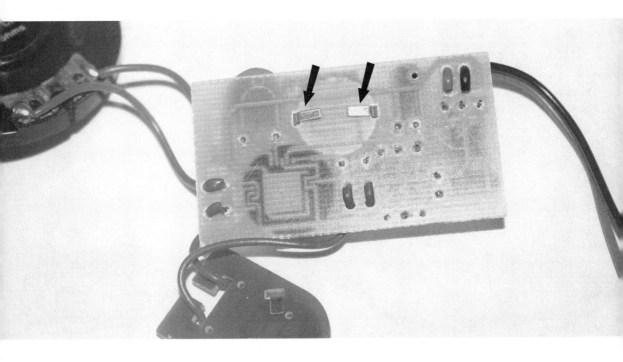

Turn the board over and pull the grey switch until it is released from the circuit board. The tabs you bent up will slip through the circuit board.

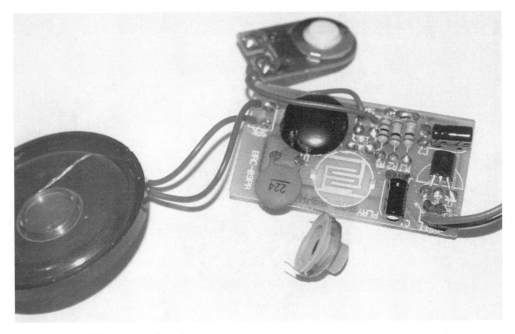

Switch removed from circuit board

Cut the two wires near the battery clip. Strip off the ends of the wire insulation, as shown.

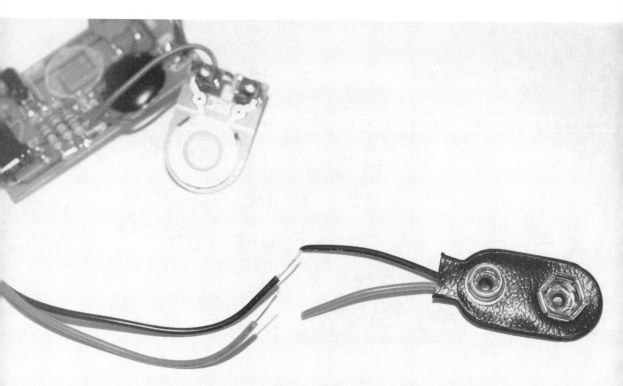

Insert the coin cell batteries into the coin cell battery holders.

Solder two 6-inch-long wire-wrap wires to the lever switch using the terminals near the lever end.

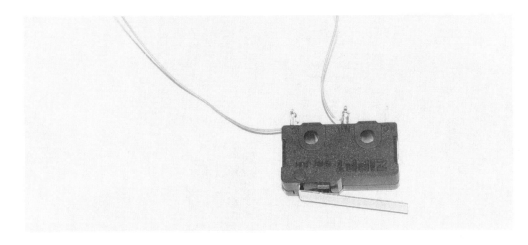

Take one of the wires (either wire) from the lever switch and solder it to the negative lead of Capacitor C1 on the circuit board.

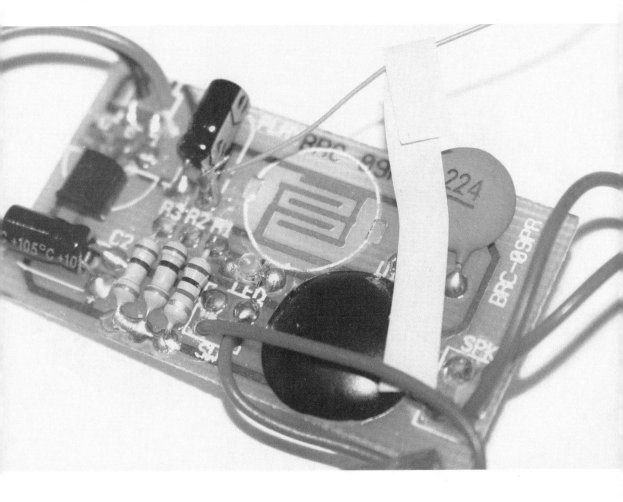

Take the other wire from the lever switch and solder it to R3 on the circuit board.

Using wire-wrap wire, connect (+) and (–) on the coin cell battery holders.

Solder the red wire from the circuit board to the free (+) on the coin battery holder.

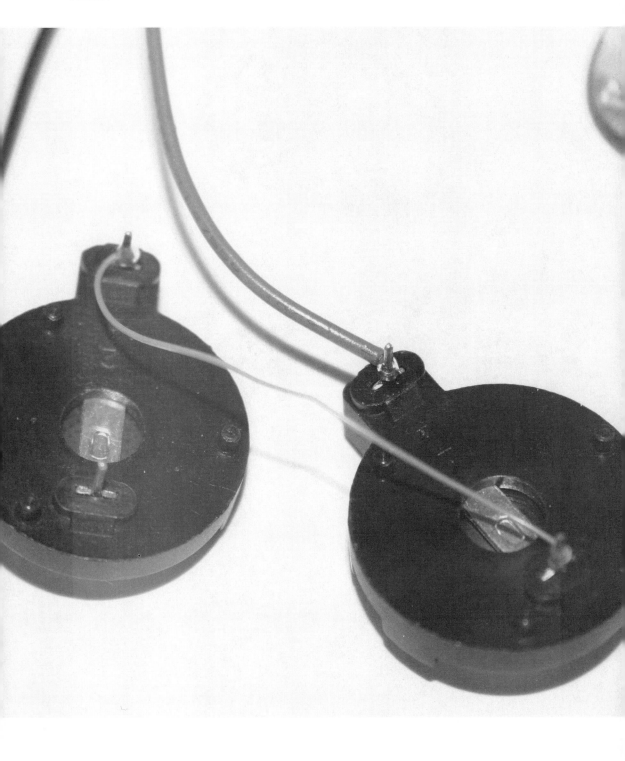

Solder the black wire from the circuit board to the free (–) on the other coin battery holder.

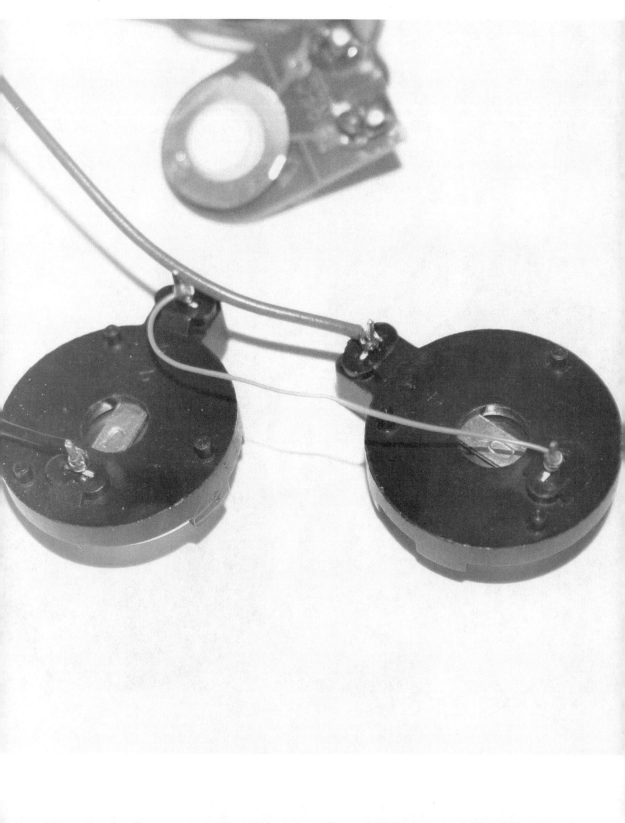

Next, make the battery holder pins as short as possible. Cut the end off each of the four battery holder pins, being careful not to cut any wires.

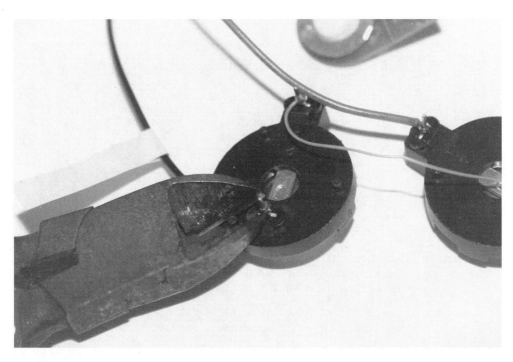

Remember the two pieces of cardboard you cut earlier? Align the holes and glue them together. The cardboard piece with the paper attached should be the lower piece, with the paper on the bottom of the lower piece.

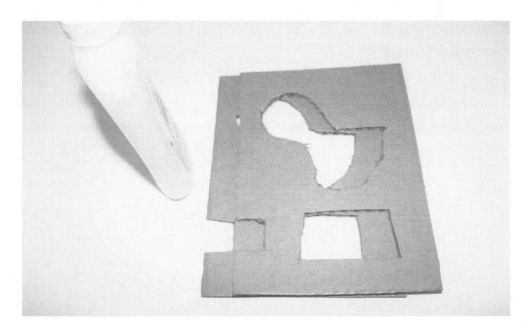

On the upper piece of cardboard (opposite the paper), cut a path from the circuit board to the battery holders. This will allow the red and black wires to "hide" inside the cardboard structure.

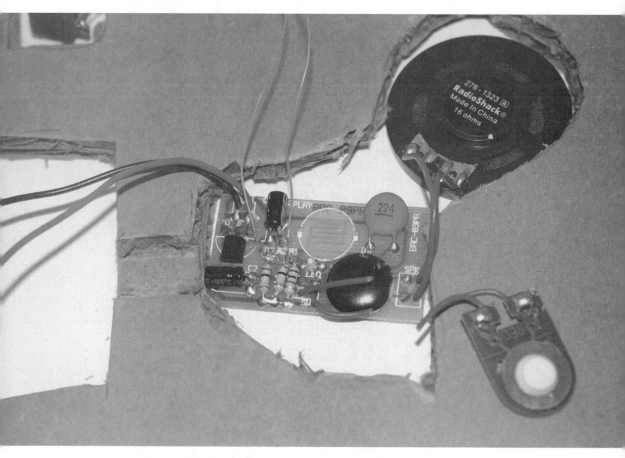

Path in cardboard and switch removal.

If the recorded message is acceptable, cut the two red wires that go to the "record" push button and remove the record push button, as shown above.

Place the circuit board, battery holder, and lever switch into the cardboard cutouts.

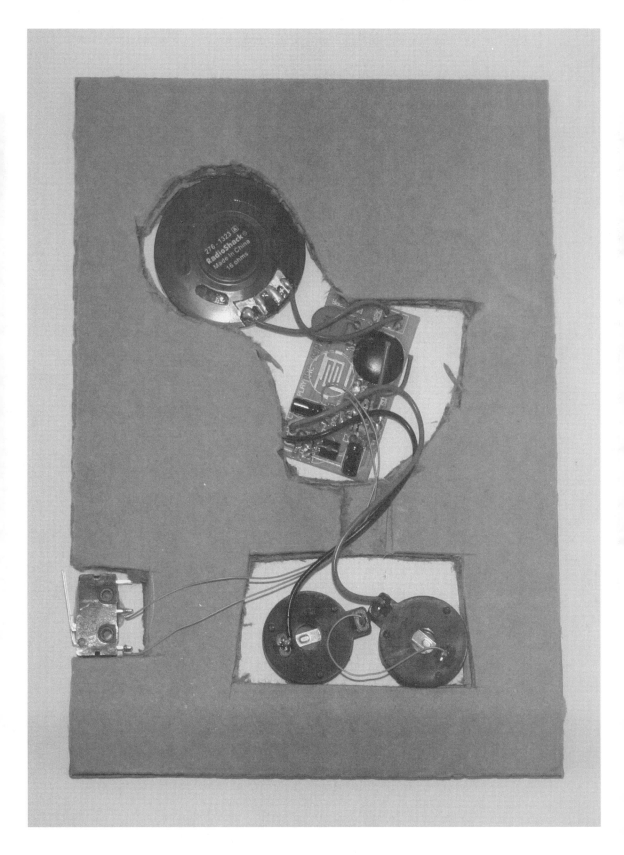

Glue the lever switch into place. The lever must protrude from the cardboard.

Use the 5¼ by 7 inch piece of cardboard for the front cover. Align the front cover so that the top, bottom, and right sides match the lower cardboard piece.

¼-inch cardboard overhang

Tape the inside "hinge" of the card. Be sure that opening the card causes the lever to be pushed. Pushing the lever causes the module to play your prerecorded message.

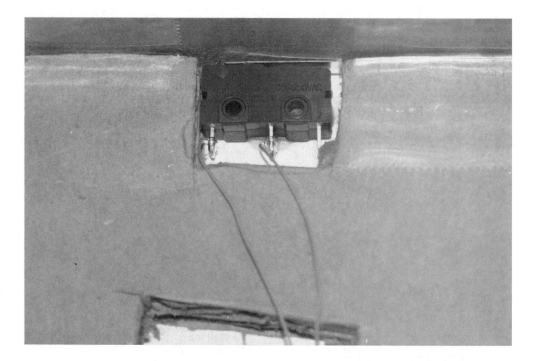

Draw or print a design on a 5¼ × 7 inch piece of paper. Glue this paper to the front face of the card.

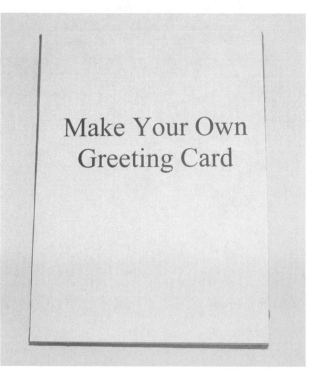

Prepare the inside design on a 5 × 7 inch piece of paper and glue it into place.

Put someone on a pedestal!

At this point, you have created a working, customized, talking greeting card. Considering all the effort and expense involved so far, it would be a shame if you didn't design something for the inside left cover and the back of the card. Get creative!

7

Soldering

The first step in soldering is to make a good mechanical connection of the wires you are soldering. The wires need to be well connected physically before solder is applied. I suggest using one of the cold soldering irons (such as Radio Shack #64-2102). This type of iron is powered by batteries, and the tip heats only when making contact with metal.

Remember to wear safety goggles. The remote chance of hot solder splashing into your eyes is not a risk you should take. Bring the soldering iron tip in contact with the connection and bring the solder in at the same time. Melt just enough solder to cover the junction and then remove the solder and the iron.

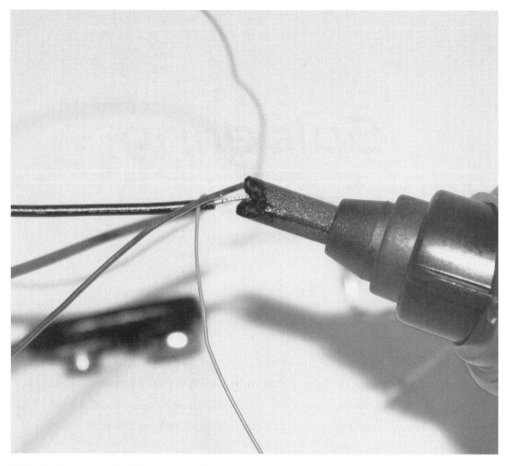

Soldering iron tip and solder at connection

Do not linger with the hot tip at the connection. Too much heat easily can destroy electronic parts. Insulation on the wires will melt and curl away. And remember, the junction and the soldering iron tip will be very hot—at least for a minute or so.

Practice a few connections before you try to solder expensive components in close quarters.

8

Smiling Picture Frame

When you approach this picture frame, the person in the picture "smiles." Use a photo of yourself or a friend. For fun, you can turn the smile into a frown, a toothless grin, or a clown's lips.

Parts List

Servo motor, www.jameco.com, #358635

(2) Wood scraps ($4^1/_2$" x $^3/_4$" x $^3/_8$")

(2) Wood scraps (3" x $^3/_4$" x $^3/_8$")

Wood base ($3^1/_8$" x 1" x $^3/_4$")

8" x 10" acrylic picture frame

Small wood screws

Old CD

Glue

8" x 10" Picture or photo

2" x 3" Paper

Tape

Velcro tape

Servo Control Module Puppet 1 Controller, Blue Point Engineering
    (www.bpesolutions.com), PCA-0010

1 K resistor

NPN switching transistor, Radio Shack, 276-1617

PIR sensor module, www.parallax.com, #555-28027

Wire-wrap wire

(4) AA rechargeable batteries

Solder

Tools List

Screwdriver

Pencil

Utility knife

Drill

$7/8''$ drill bit

Wire wrap tool

Soldering iron

Start this project by using small wood screws to attach a servo motor to two
$4\frac{1}{2} \times \frac{3}{4} \times \frac{3}{8}$ inch strips of wood, as shown, to make a servo motor holder.

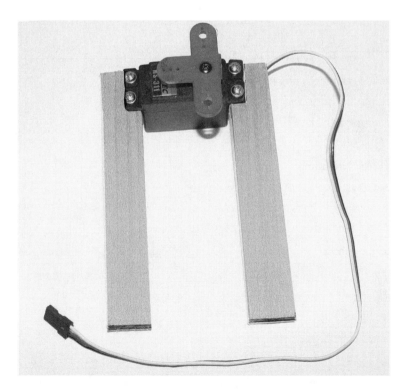

Use small wood screws to attach the wood base (3¹/₈″ × 1″ × ³/₄″) to the other wood pieces (3″ × ³/₄″ × ³/₈″). The 3¹/₈ inch piece of wood shown below was cut at a 30-degree angle to match the angle of the acrylic picture frame.

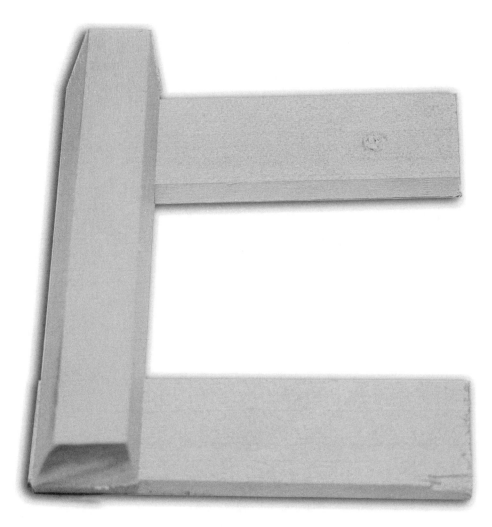

Wood base

Screw the servo motor holder to the base to create a servo assembly.

Servo assembly

Glue a CD to the servo assembly, taking care that the shaft of the servo motor is precisely in the center of the CD's hole.

Tape the picture to the back of the frame, then set the servo assembly in place. The CD should be as close to the picture as possible, but it should not touch the picture when it rotates.

Using a pencil, lightly trace the edge of the CD onto the back of the picture.

Shine a light through the picture (from the front) and mark the area you want to modify on the back of the picture. The area should rest entirely within the circle drawn on the back. Then, using a utility knife, cut out the area of interest.

Next, use a pencil to outline the area of interest onto a sheet of paper. This area is the spot where you will modify the picture by drawing teeth, adding lipstick, pasting a frown, or whatever you like.

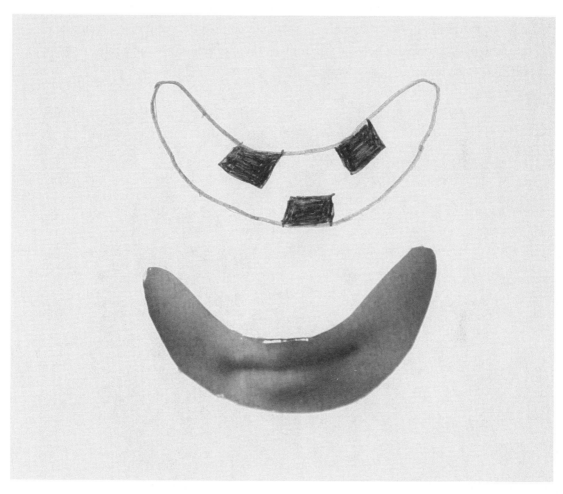

Area of interest and replacement smile

With the servo motor in its resting position, tape the original smile (the one you cut out) in place. Resting position on the servo motor is the spot where the servo "rests" when power is applied to the servo control module. This normally involves turning the motor clockwise (viewed from the back) until it reaches a stop. Servo motors do not spin, they turn about one revolution and can be made to stop anywhere within that revolution.

Tape the replacement smile onto the CD in a different place, such that rotating the CD will cause the replacement to fit where the original smile was. You don't have to worry about exactly where on the CD you tape the second smile, but it must be placed so that when the CD rotates, the replacement smile aligns with the original smile hole.

Original and replacement smiles

The picture, when placed in the acrylic frame, will have a hole where the original and replacement smiles must fit.

Drill a ⅞-inch hole in the lower part of the frame to accept the sensor module.

Hole with sensor module in place

Cut a hole in the picture to correspond to the sensor module hole.

Glue the servo assembly to the base of the picture frame.

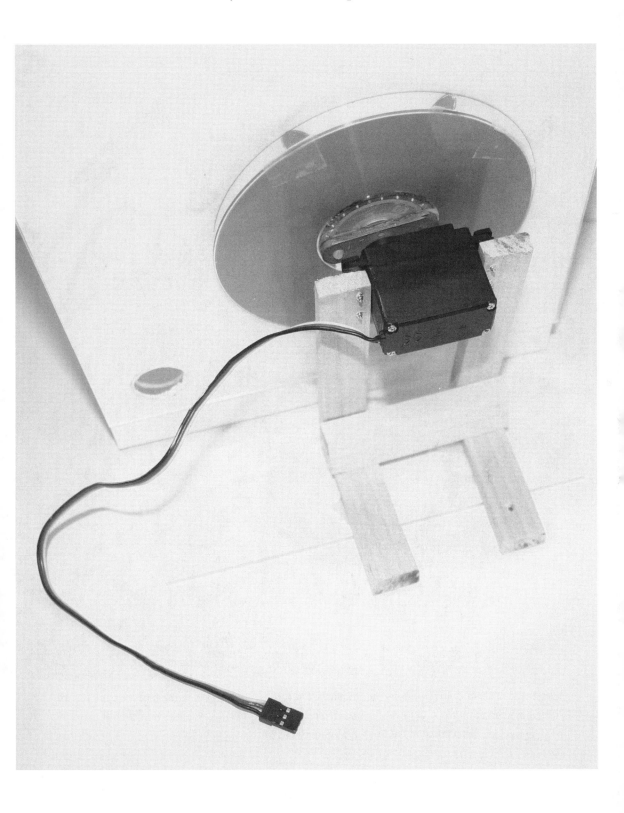

On the servo control module, wrap a wire-wrap wire around the r/p pins as shown.

Wire on r/p of servo control module. The r/p pins are "record" and "play." When the pins are connected, the servo control module is prepared to record. When disconnected, the servo control module will play the action that has been recorded.

The connection you just made is temporary. It will be used to program the module to rotate exactly where you want the smiles to be. After that programming is complete, you will remove this wire.

Place four rechargeable AA batteries in the AA battery holder. Solder the battery holder leads to the wire-wrap wire. Bring the positive wire (the wire-wrap wire attached to the red battery lead) to the (+) position on the servo module. Bring the negative wire (the wire-wrap wire attached to the black battery lead) to the (–) position on the servo module. Notice that the servo module has "5 VDC" printed beneath the power pins. If you use alkaline batteries instead of rechargeable AA batteries, you'll wind up with six volts DC. Six volts might work, but it could damage the circuitry or cause erratic operation. This is the type of detail that can cause projects to act strangely or fail prematurely.

Attach the servo motor to the module, with the yellow wire at the end with the "y."

Apply power to the servo control module.

The servo motor will go to its "rest" position. The original smile should be lined up with the hole in the picture. Press the blue button. While holding the blue button down, turn the black wheel until the replacement smile lines up with the hole in the picture. Then turn the black wheel until the original smile is back in place. Release the blue button. Programming is now complete. Remove the wire-wrap wire from the r/p pins on the servo control module. Press the blue button. the original smile will be replaced, then come back. Remove the batteries.

Now it's time to make the sensor module "push the blue button" whenever someone approaches. First, take a piece of wire-wrap wire and connect the center lead of the transistor to one end (either end) of the 1K ohm resistor.

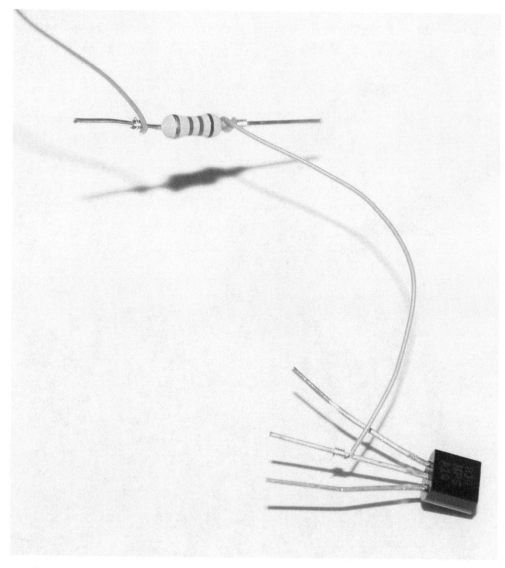

Attach transistor to resistor.

Next, attach the "out" pin of the sensor module to the 1K resistor (attaches to the free end of the resistor) with wire-wrap wire.

Using wire-wrap wire, connect the emitter of the transistor to (–) on the servo control module. If you are looking at the flat side of this transistor, the emitter is the pin on the left.

Transistor emitter connects to (–).

Using wire-wrap wire, connect the collector of the transistor to the pin closest to the word *go* on the servo control module. Looking at the flat side of this transistor, the collector is the pin on the right.

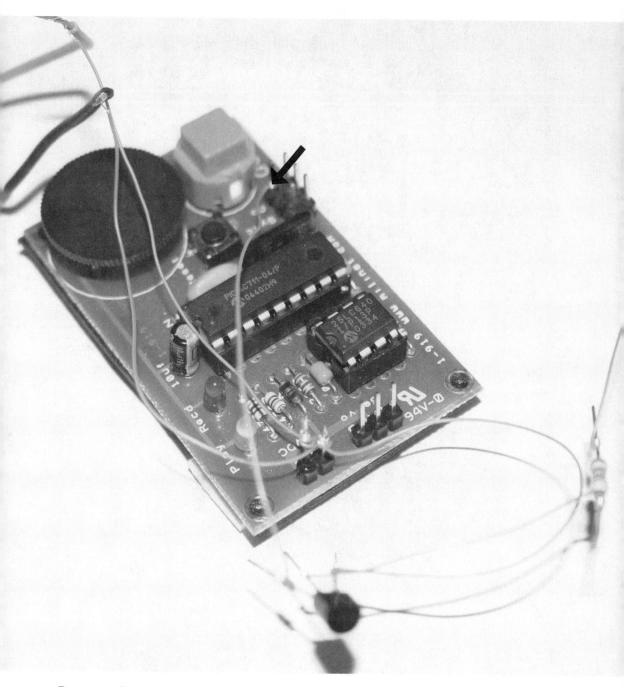

Transistor collector connected to *go*

Attach (+) from the battery to (+) on the sensor module. Connect (−) from the battery to (−) on the sensor module. In the upper left-hand corner of the sensor module is a jumper. That jumper should connect the top pin and the center pin. This causes the module to "press the blue button" only once when a person approaches. In the other position, the "press the blue button" action will repeat until the person leaves.

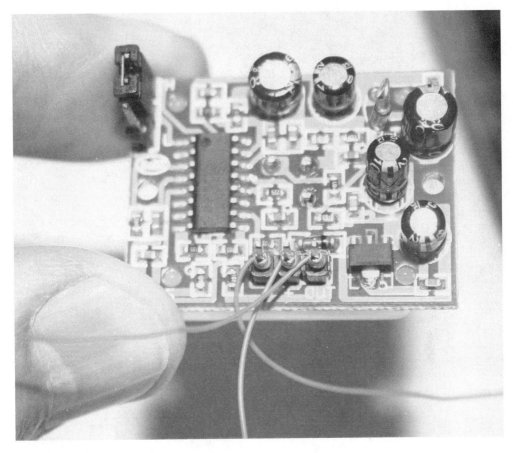

Power connection to sensor

Now is a good time to test the project. Place the batteries in the battery holder. Step away from the project for about two minutes—the sensor needs time to adjust. Then stand in front of the sensor. The alternate smile should appear.

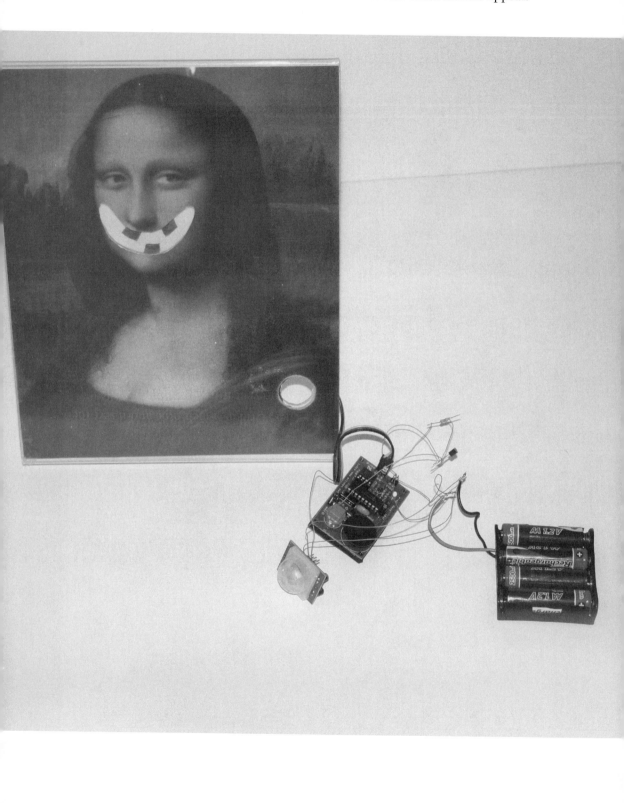

If all is well, glue the sensor into the frame. Attach the battery holder to the picture frame (Velcro tape was used here). Attach the servo control module to the servo assembly (glue or Velcro). Use electrical tape to cover any exposed wires that might touch any metal parts. For maximum reliability, cover the transistor and resistor with tape. If you are striving for the most complicated look, let the transistor and resistor hang freely.

Circuitry secured to frame

Attach the word *Smile* to the front of the frame.

Approach the frame. The alternate smile should appear.

9

Flashlight Without Batteries

In this project, you will make a flashlight that works without batteries. Even more amazing, you can recharge it in three minutes and it will run for more than 24 hours.

Parts List

(2) 220 farad capacitors, www.digikey.com, #589-1013-nd

Insulated wire, black and red

Solder

LED, high brightness, www.jameco.com, #217525

12" x 12" acrylic plastic sheet, $1/8$" thick

Permanent marker

(2) C clamps

Epoxy

$1/4$" jack, Radio Shack, #374-280

Metallic tape

SPST rocker switch, Radio Shack, #275-693

Glue

Electrical tape

Tools List

Wire cutters
Soldering iron
Scoring knife (for plastic)
Single-hole paper punch
3-volt DC power supply
Drill
$\frac{1}{4}''$, $\frac{1}{16}''$, and $\frac{3}{4}''$ drill bits
Metal straightedge

First, solder a wire from the (+) on one capacitor to the (–) on the other capacitor.

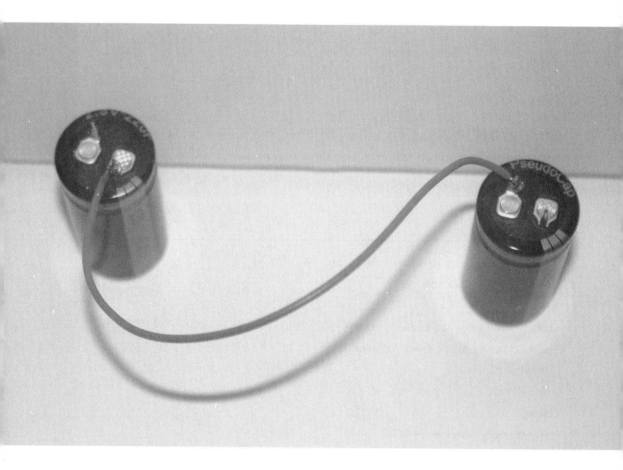

Solder a 4-inch length of red wire to the unused (+) terminal on the capacitor. You will call this capacitor wire (+).

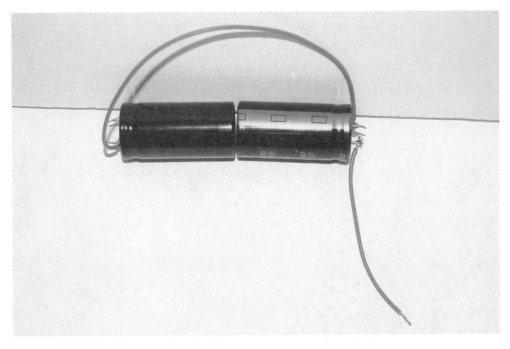

Solder wire to (+).

Solder a 2-inch length of black wire to the unused (−) terminal on the capacitor. You will call this capacitor wire (−).

Solder wire to (−).

Next cut the ⅛-inch plastic sheet for the flashlight case. Generally, this is accomplished by scoring the plastic, then breaking it along the scored lines. But first, mark the plastic with a marker, then use the C clamp to hold a metal straight-edge next to the mark. Then, use the tip of a scoring knife (usually sold next to the plastic sheets) to create a fine groove in the plastic along the line you marked.

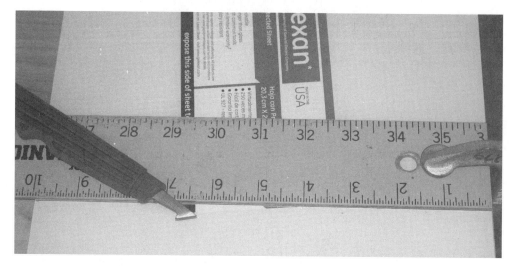

Correct method to cut plastic

Using the scoring edge (not the knife tip), pull the knife along the fine line several times until you create a groove. The plastic can now be snapped along the score line.

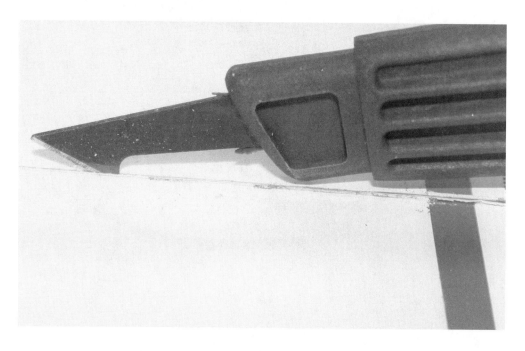

Practice this method a few times until you master the technique.

Now you're ready to build your flashlight case. First, cut out three 1¼ × 7 inch pieces of plastic and glue them together with epoxy. This is the flashlight body.

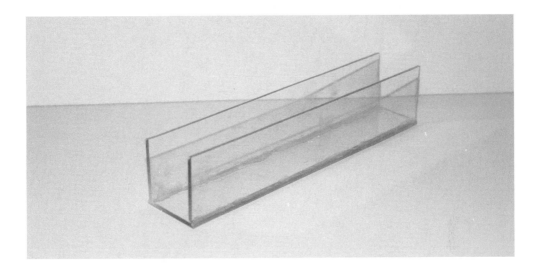

Cut a plastic piece 1¼ inches by 1½ inches. Drill a ¼-inch hole and insert the ¼-inch jack. Hold the jack with one hand (to prevent it from rotating) while threading the nut and washer.

Solder a 2-inch-long black wire to the end of the jack near the tab. Solder a 3-inch-long red wire to the other side of the jack (away from the tab).

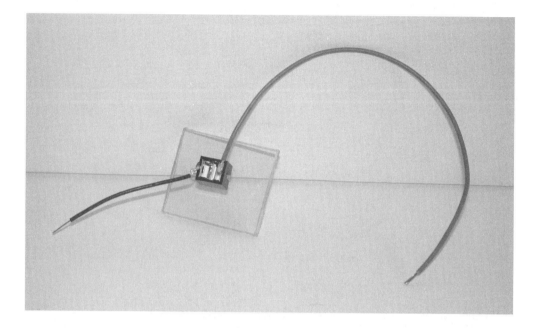

Solder the red wire from the jack to the capacitor (+). Solder the black wire from the jack to the capacitor (–). These wires form the path that will be used to charge the capacitors.

Cut a piece of plastic, 1 × 1¼ inches. Drill two ¹⁄₁₆-inch holes in the center, side by side, the same distance apart as the leads on the LED.

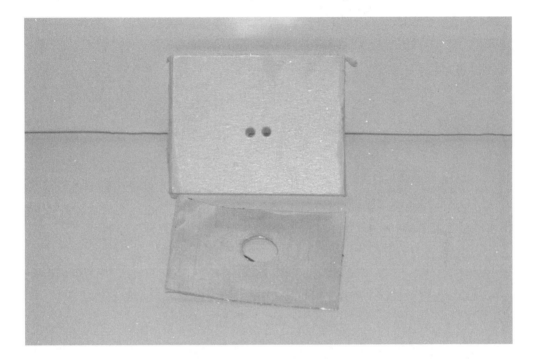

Cut a piece of metallic tape, 1 × 1¼ inches. Punch a hole in the center of this tape using a hole punch. Remove the paper backing from the tape and fasten the tape to the plastic. This is part of the reflector for the flashlight. Insert the LED leads through the holes.

Cut a piece of plastic 7 × 1¼ inches. Drill a ¾-inch hole 2¼ inches from one end. This is the plastic switch holder. Press the on/off switch through the hole and secure it by threading the switch's plastic nut.

Solder one end of a 6-inch length of black wire to one end to the capacitor (−). Solder the other end to the on/off switch.

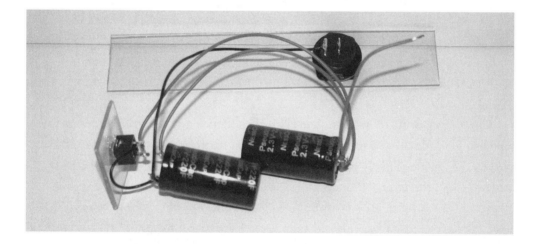

Solder the wire from capacitor (+) to the long lead of the LED.

Cut a 2-inch length of red wire. Solder one end to the short lead on the LED. Solder the other end to the on/off switch.

When the switch is on, power will flow from capacitor (+) through the LED, through the switch, and back to capacitor (−). This will cause the LED to light up. If everything is working properly, carefully place the capacitors and the jack into the flashlight body. Glue the jack piece onto the end of the flashlight body.

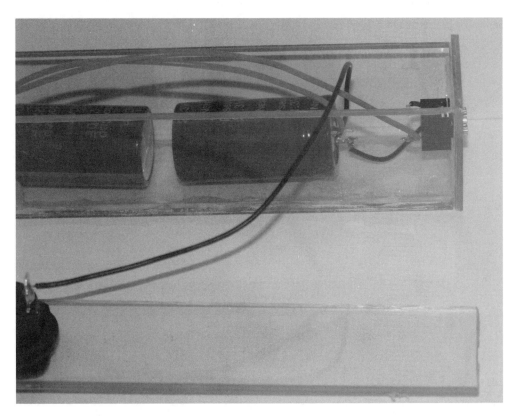

Glue jack piece to flashlight body.

Cut a 1 × ¾ inch piece of plastic. Insert it into the flashlight body, between the jack and the capacitor. This will prevent the capacitor from sliding into the jack, thus avoiding a potential short circuit.

Separated jack and capacitor

Place electrical tape over the on/off switch wires.

Glue the LED holder into the flashlight body.

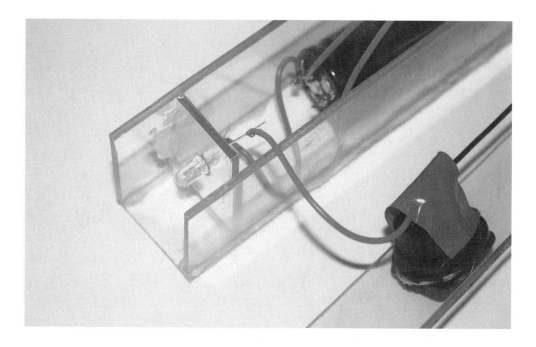

Place metallic tape inside the flashlight body and on the plastic switch holder to form the reflector.

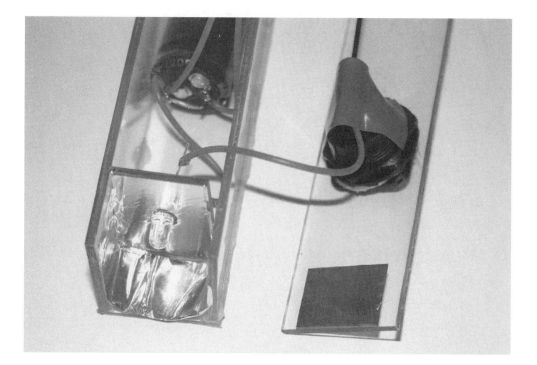

Glue the plastic switch holder to the flashlight body. Cut a 1½ × 1¼ inch piece of plastic for the lens. Glue the lens in front of the LED.

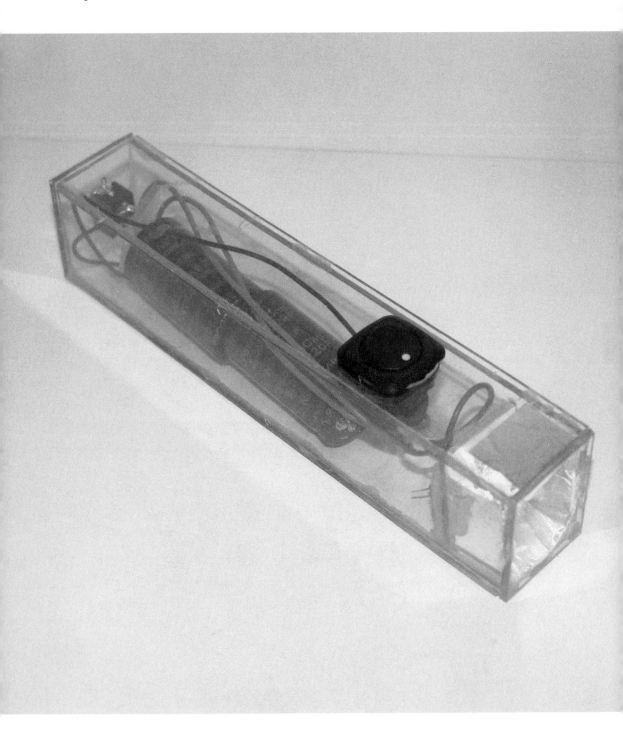

Plug a 3-volt DC power supply, (+) tip, into the back for three minutes. Remove the power supply and enjoy your new flashlight!

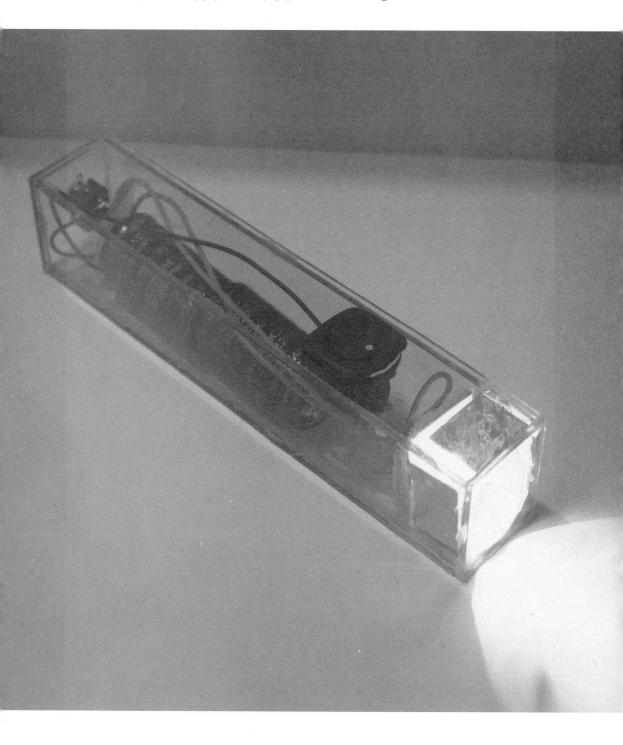

10

Transistor Switch

For these types of projects, it is helpful to know a little bit about transistors. Most integrated circuits (and little modules) do not produce output signals that are powerful enough to do anything practical. You sometimes want to take a tiny signal and turn on a light, start a motor, or control a relay.

Rather than dealing in-depth with transistors, let's look at how to use a plain, inexpensive NPN transistor. NPN refers to a type of transistor where (+) is used to turn the transistor on. Think of a transistor as sort of an electrical lever—put a little bit of energy in and you can control a lot of energy.

When you look at transistor packaging, the manufacturers will often tell you which of the three legs is the base, which is the collector, and which is the emitter. For the NPN transistor, the emitter goes to (–), the base is the control pin, and the collector is where (+) comes in.

Using an NPN transistor to switch power, you can make a "touch" switch to turn an LED on. Look at the photo on page 120. When the transistor is turned on, power will come through the resistor and LED, and the LED will be illuminated.

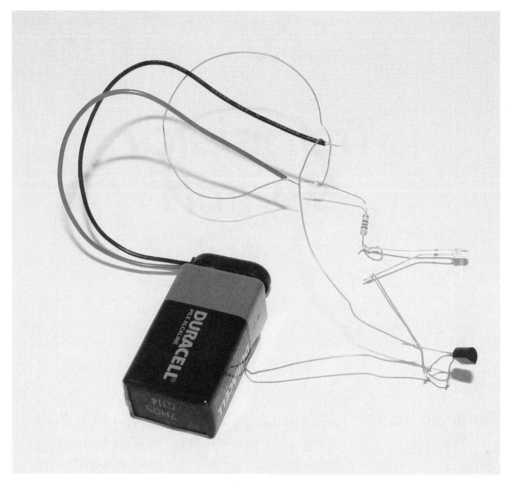

LED switch circuitry

By touching the base of the transistor with one finger and battery (+) with the other, a tiny amount of power flows through your skin and turns the transistor on. Any small current through the base will turn the transistor on.

Very little power will flow through a person's skin from a 9-volt battery. If you are using something else to turn the transistor on, to prevent too much power from going through the base of the transistor, you can connect a 1K ohm resistor to the base and then attach your (+) power source to the other end of the resistor.

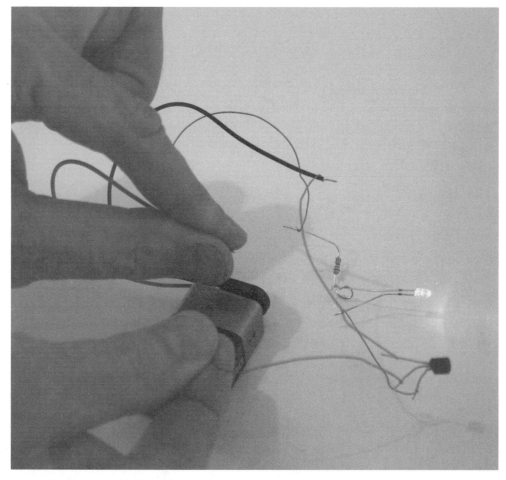

LED touch switch on

You can replace the LED and resistor with a relay and let the relay control a motor or buzzer.

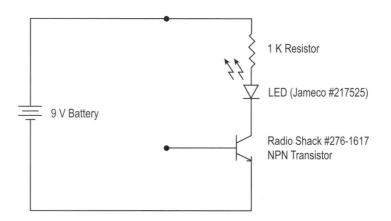

11

No-Battery Electric Car

In this project, you will build a simple electric car that is not powered by batteries. Instead of using a rechargeable battery, this car is powered by "super" capacitors.

Electric car technology today faces two major problems. First, the batteries take too long to charge. You cannot take an electric car to a "filling" station and recharge quickly—it would take hours. Second, rechargeable batteries can be quite costly and tend to wear out after a couple of years or a few hundred charge/discharge cycles.

Capacitors, on the other hand, can be charged in minutes rather than hours, and they can take millions of charge/discharge cycles. So why, you may ask, aren't capacitor-powered electric cars all over the roads, eliminating the need for gasoline? There is no conspiracy; capacitors have problems too. A capacitor weighs much more than a battery of similar capacity. High-density capacitors (which are still heavier than batteries) cost much more for the same energy storage capacity. Last, capacitors have a peculiar way of releasing their energy. In a capacitor, voltage declines in a straight line over time—100 volts, 75 volts, 50 volts, 25 volts, 0 volts. In a battery, voltage tends to stay in a usable range for most of its life—100 volts, 100 volts, 99 volts, 90 volts, 0 volts.

When someone develops a super–high-storage capacitor at a reasonable cost, the energy usage patterns of the world will be able to shift dramatically. In this chapter, you will explore the principle of making a vehicle powered by a capacitor.

Parts List

(4) ¼"-thick plywood pieces (for frames and side rails, see page 139)

½"-thick plywood piece (for base, see page 139)

Glue

(2) Wooden dowels, 5¾" long, ⅜" diameter

(4) Wooden bearing holders, 1¾" x 1¼" with ½"-diameter hole in center

(4) Bearings ⅜" inside diameter, ⅞" outside diameter, www.vxb.com, #R6zz

Small wood screws

(4) Used CDs

Motor, 6V, 6 rpm, www.jameco.com, #253323

(2) Grommets, ⁷⁄₁₆" inside diameter, ¾" outside diameter

(1) Grommet, ¼" inside diameter, ⁷⁄₁₆" outside diameter

(1) Grommet, ¼" inside diameter, ½" outside diameter

Sprocket, plastic chain, www.jameco.com, #157075

Sprocket, ⅜", 30 teeth, www.jameco.com, #400557

Sprocket, ⅜", 48 teeth, www.jameco.com, #400590

Insulated wire

SPST toggle switch, Radio Shack, #275-624

(4) Capacitors, 1 farad, 5.5 volt, www.jameco.com, #142957

Power cord, Radio Shack, #273-1742

Power supply, 9-volt 800-milliamp, Radio Shack, #273-1768

Tools List

Jigsaw

Screwdriver

Drill

½" drill bit

Hammer

Using the templates on page 139, cut out the base, frame, and side rails from the plywood, using ¼-inch-thick plywood for the frame and side rails, and ½-inch-thick-plywood for the base. Attach the base to one of the frame pieces with glue.

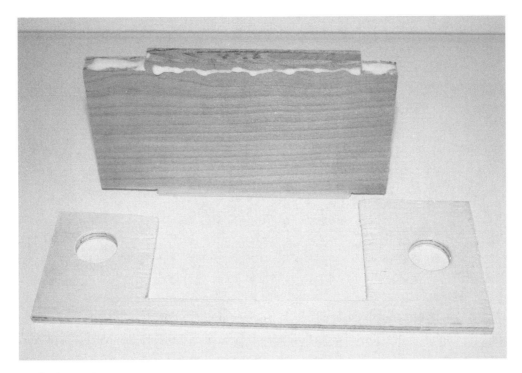

Apply glue as shown.

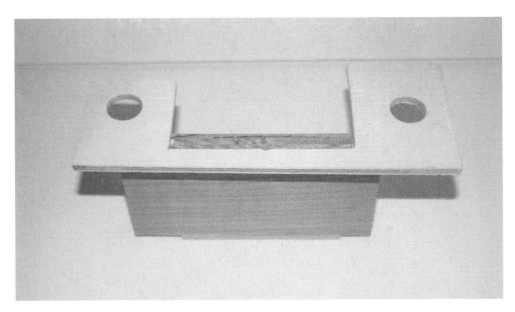

Attach the base to the frame.

To further secure the base to the frame, drill pilot holes—holes slightly smaller in diameter than the wood screw threads—and screw the base to the frame.

Using a hammer, drive the 5¾-inch wooden dowel, which will be the car's axle, through the bearings. Be certain that two bearing holders are placed between the bearings.

The bearings should each be 1 inch from the end of the axle.

Bearing and bearing holders

Place one end of the bearing/axle assembly into the frame, as shown.

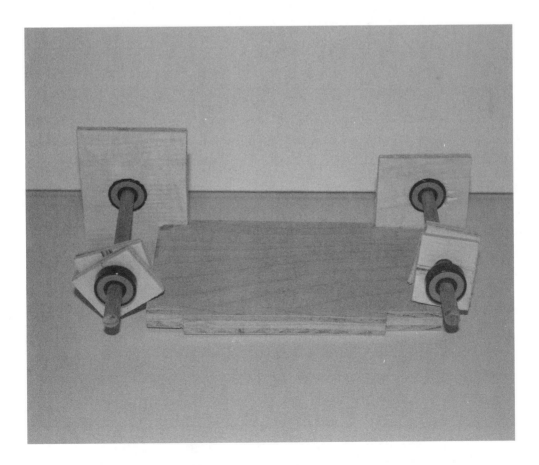

Slide a bearing holder up against the frame.

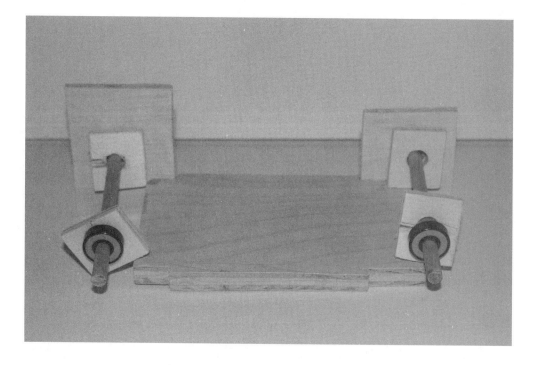

Fasten the bearing holder with screws.

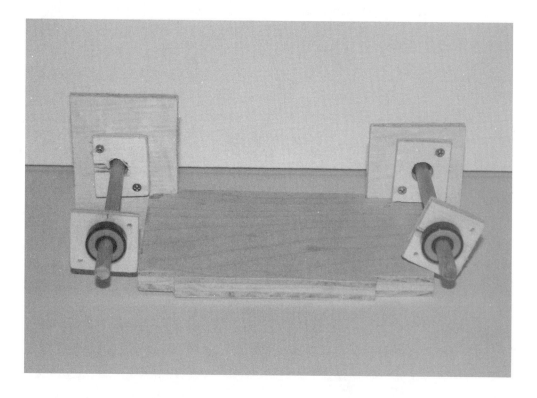

Glue the other frame piece into place, making sure that the bearing is seated in the frame. Secure the frame piece with screws. Fasten the side rail with glue and screws. Glue bearing holders to the outside of the bearing/axle assembly.

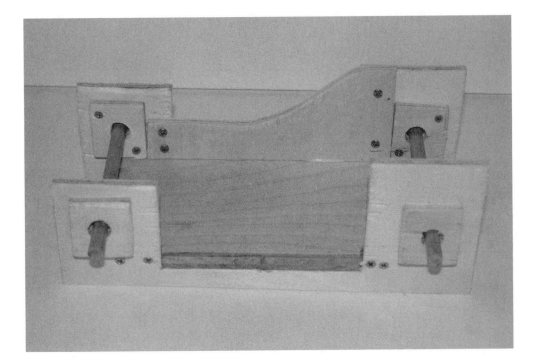

Fasten the motor to the not-yet-attached side rail with screws.

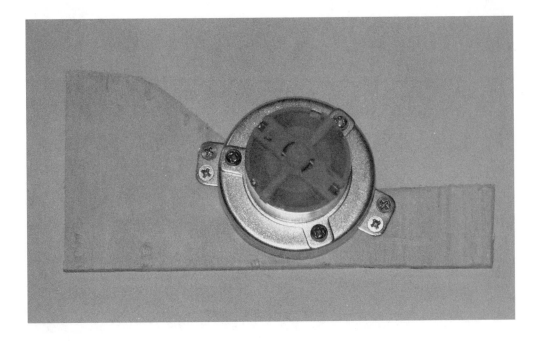

Secure this motor side rail to the frame with screws. Note that the motor side rail is fastened to the outside of the frame, while the other side rail is inside the frame.

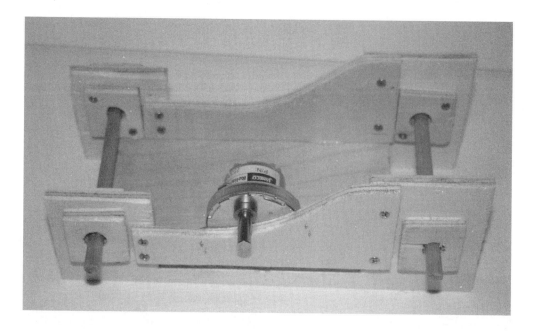

Wrap masking tape around the end of the axles.

Take two CDs and install a grommet (⁷⁄₁₆-inch inside diameter, ¾-inch outside diameter) in the center of each. Slide the CDs onto the front axle.

Take the small grommet (¼-inch inside diameter, ⁷⁄₁₆-inch outside diameter) and cut off one side.

Insert the cut-off side of the grommet into the 48-tooth sprocket. Then push the sprocket/grommet onto the motor shaft. Put the chain on the sprocket.

Cut off one side of another grommet and insert it into the 24-tooth sprocket. Place the smaller sprocket/grommet onto the rear axle. Secure the chain.

Place two CD wheels on the rear axles. In the side rail opposite the motor, drill a ½-inch-diameter hole. Insert a grommet (¼-inch inside diameter, ½-inch outside diameter) into the hole.

Place a switch inside the grommet.

Solder one end of a 3-inch length of insulated wire to a motor terminal.

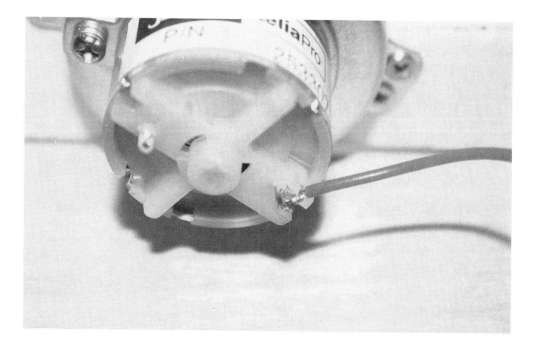

Solder the other end of the wire to a switch terminal.

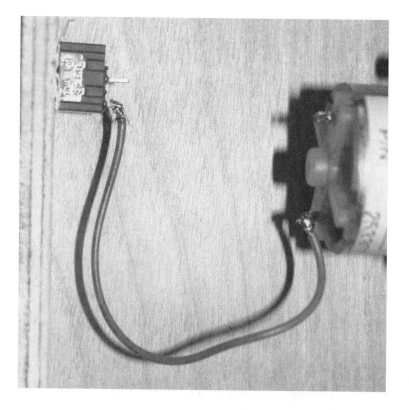

Connect a wire from (–) on one capacitor to (+) on another. Solder a 2-inch length of wire to the empty (+) terminal.

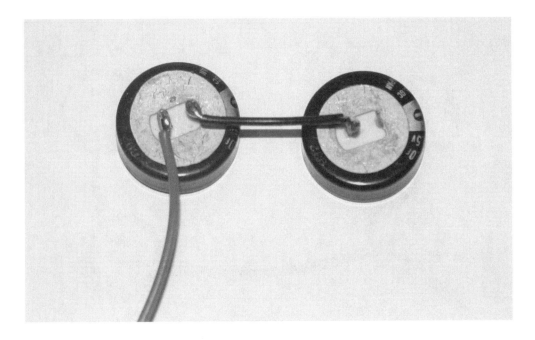

Solder a 2-inch length of wire to the empty (–) terminal.

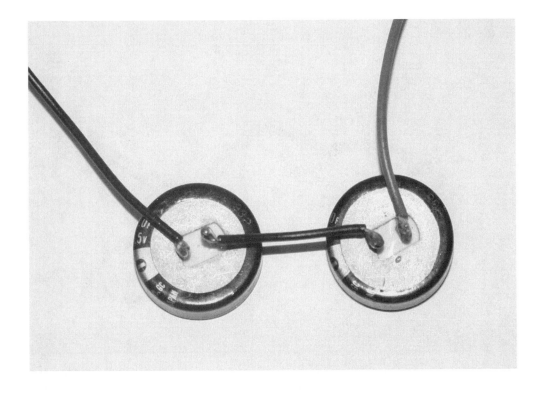

Connect (+) to (−) on the other capacitor pair. Then connect (+) on one capacitor pair to (+) on the other set. Connect the (−) terminals together.

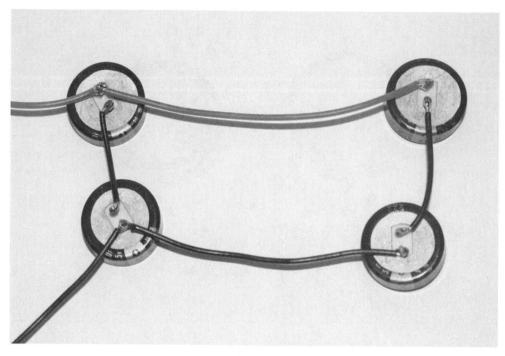

Four capacitors connected together

Connect the power cord to the capacitors, (+) to (+) and (−) to (−).

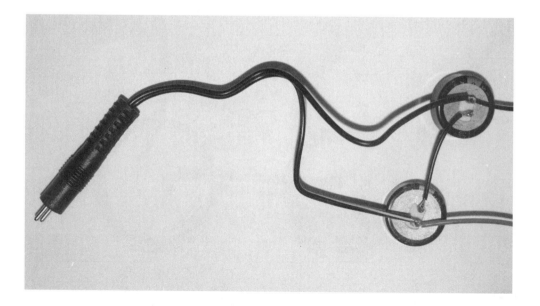

Connect the (+) lead from the capacitors to the switch.

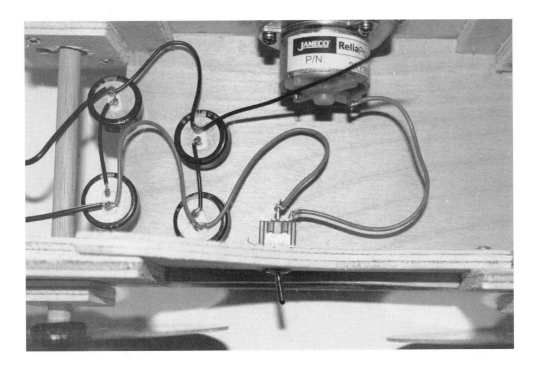

Connect the (–) lead from the capacitors to the motor.

Your car is now complete. Plug in the power for about three minutes. Remove the power supply. Turn the switch on, and the car will move.

You can paint or decorate the car to suit your taste.

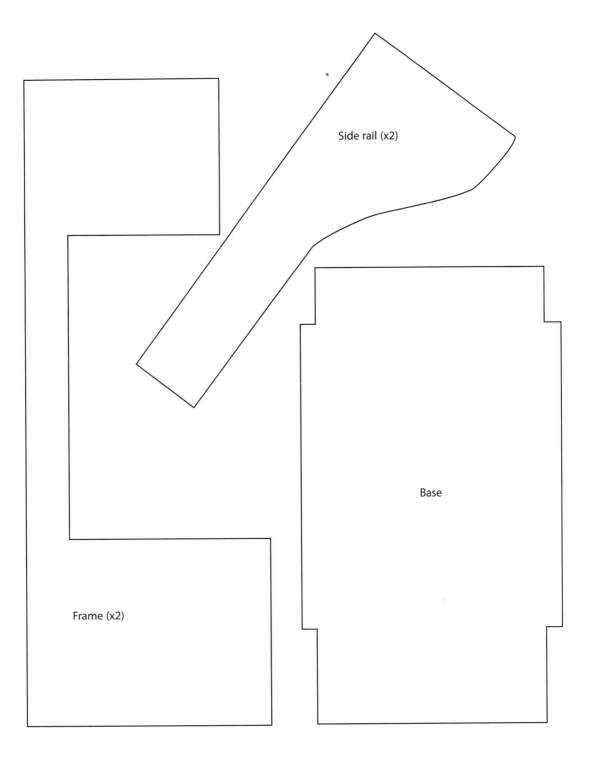

Side rail (x2)

Base

Frame (x2)

Building a Larger Car

To build a larger car, two major questions come to mind: How much capacitance will I need? How long will the car run?

To come up with an estimate, you have to consider the physical characteristics of the capacitors and use a few equations. Ultra capacitors are rated for low voltage—2.3 volts is typical. Assuming that you want to power a 6-volt motor, you will have to place three capacitors in series, (+) connected to (−) on each capacitor. This will allow you to place up to 6.9 volts on your capacitors. When you add capacitors in series, the total capacitor value is equal to the reciprocal of the individual capacitors. $1/C = 1/C_1 + 1/C_2 + 1/C_3$. For our example, if you place three 220-farad capacitors in series, you will have one 73.3-farad capacitor that can handle 6.9 volts.

Now assume that your motor draws 1 amp at 6 volts. This ignores certain facts. The motor will draw more than 1 amp as it starts up, and it will draw more than 1 amp if the load it is turning is increased. Voltage divided by current equals resistance ($V/I = R$). Considering your simple motor example, 6 volts divided by 1 amp yields an effective resistance of 6 ohms.

The time constant for a capacitor—how long it will last until it is two-thirds depleted—is approximately equal to resistance multiplied by capacitance ($T = RC$). Time is measured in seconds. For your example, the motor above will run for RC seconds: 6 x 73.3; about 440 seconds.

One problem is that the motor will move slower and slower as time goes on, as the capacitor voltage steadily drops. Without additional circuitry to steady the voltage, this is inevitable. The second problem is that the motor will use extra amps when starting up or carrying an extra load, thus decreasing the run time.

Since you are discharging the capacitor steadily at 1 amp for 440 seconds, recharging the capacitor quickly will take a high-capacity power supply. Suppose you want to recharge the capacitor in 22 seconds. Because this is 20 times as fast as the discharge rate, you will need to supply 20 amps to charge the capacitor quickly. Don't forget that your wire size will have to be much larger for the 20-amp charging current than for the 1-amp discharge current.

12

Motors

Although there are many types of motors, the ones that are easiest to use for simple projects are DC (direct current—meaning battery operated) gearmotors. A DC motor without gears is inexpensive but turns at a very high rpm and has very little power.

The nice thing about DC motors is that they are easy to reverse. Power to the motor is supplied through two pins. Put (+) on one pin and (–) on the other, and the motor will turn. Reverse the connection, and the motor will rotate in the opposite direction.

To control the speed of a DC motor, change the voltage. Lower voltage gives you a slower speed. Higher voltage gives a higher speed.

13

First-Answer Box

In some games, you need to know which player is first at answering a question. The following project can help you avoid disputes.

Parts List

Perfboard, 4" x 4", Radio Shack, #276-1395

(4) 4-pole double-throw relays, www.jameco.com, #282301

Solder

Insulated wire, red and black

(4) Push-button switches, www.jameco.com, #315432cj

Toggle switch, www.jameco.com, #317463

(4) Green LED assemblies, Radio Shack, #276-069b

12-volt DC power supply, Radio Shack, #273-029

SPST toggle switch, Radio Shack, #275-634

Box, 7" x 5" x 3", Radio Shack, #270-1807

(4) Boxes, 2" x 4" x 1", Radio Shack, #270-1802

Connector, 6-pin female interlocking, Radio Shack, #274-236

Connector, 6-pin male interlocking, Radio Shack, #274-226

Barrier strip, 8-position European style, Radio Shack, #274-678

Velcro tape

Tools List

Soldering iron

Drill

$1/4$", $3/8$", and $1/2$" drill bits

Screwdriver

Wire stripper, Radio Shack, #64-2982

This First-Answer Box has four buttons, each corresponding to a light. Pressing Push Button A will turn on Light A while locking the other buttons out. Opening and closing the reset switch turns off all the lights so that the box is ready to receive the next push of a button.

Because this project uses relay logic for operation (instead of integrated circuits), every step of the operation is straightforward. Examine the wiring diagram on page 146. First, close the reset switch. All the LEDs are off. If you push PBA (Push Button A), power flows through contact PBA, through Contacts D1, C1, and B1, through Relay Coil A, through Resistor A (RA) and LED A, causing LED A to turn on. When Relay Coil A is energized, it causes Contact A1 to close while causing Contacts A2, A3, and A4 to open. Contact A1 "seals" Relay Coil A into the energized state. When you release PBA, it has no effect. Power now flows through Contacts A1, D1, C1, and B1 to keep Relay A and LED A energized. Relay B cannot be energized, because Contact A2 is open. Relay C cannot be energized because Contact A3 is open. Relay D cannot be energized because Contact A4 is open.

To reset the system, open the reset switch. This removes the energy from Relay A and LED A, so all the lights are out. The first button pushed will energize its respective relay and LED while locking the others out.

The resistors (RA, RB, RC, RD) are used to reduce the voltage across the LED. The 12 volts needed to energize the relay would instantly burn out the LED, since the LED requires about 3 volts.

Now it's time to start building. First, cut a 4 × 4 inch piece of Perfboard.

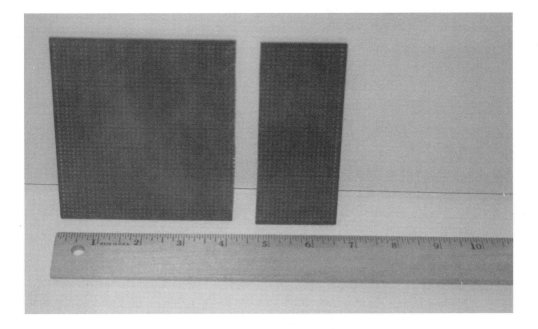

Glue four relays to the Perfboard. Label the relays A, B, C, and D, as shown.

To build this circuit, you will be connecting various points on the four relays. For point numbers, refer to the following schematic.

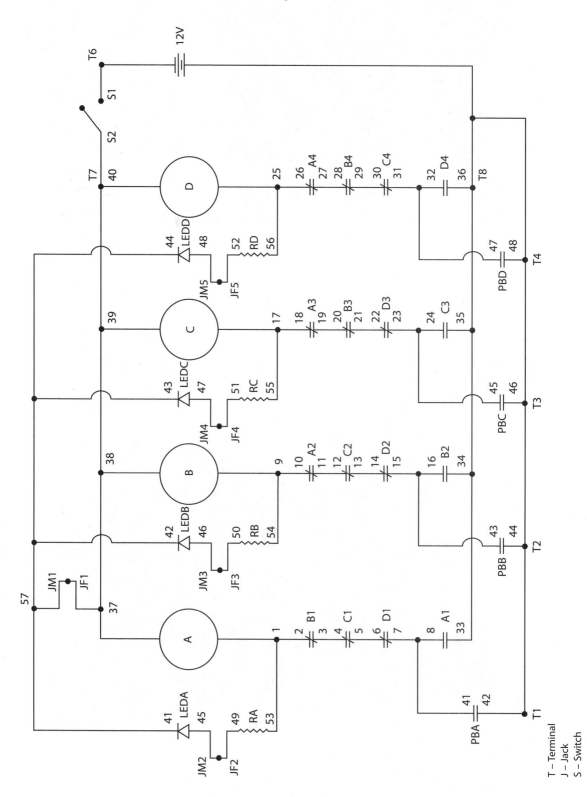

Begin by soldering a wire from Point 1 to Point 2.

Connect Point 3 to Point 4.

Connect 5 to 6.

Connect 7 to 8.

Connect 9 to 10.

Connect 11 to 12.

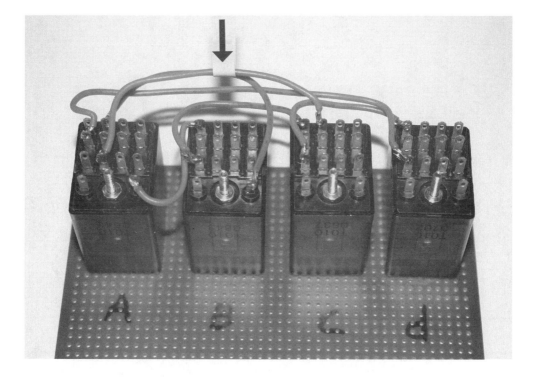

Connect 13 to 14.

Connect 15 to 16.

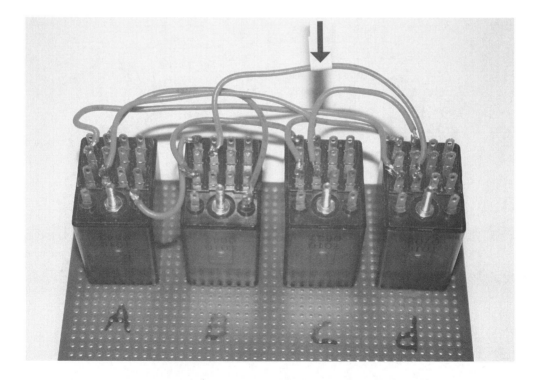

Connect 17 to 18.

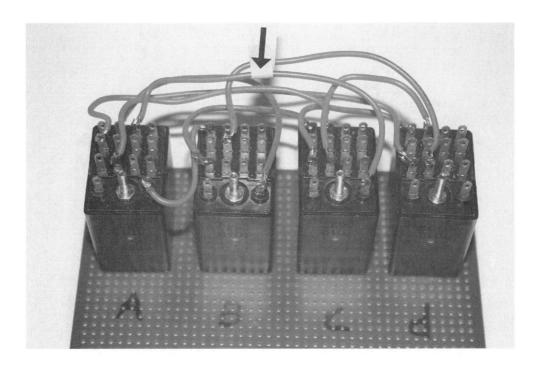

Connect 19 to 20.

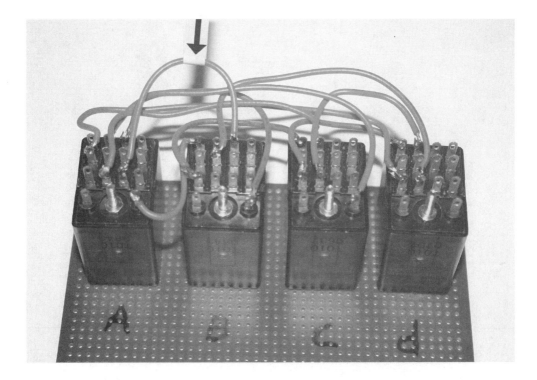

Connect 21 to 22.

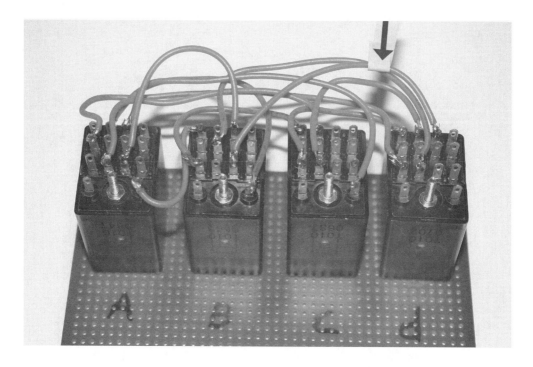

Connect 23 to 24.

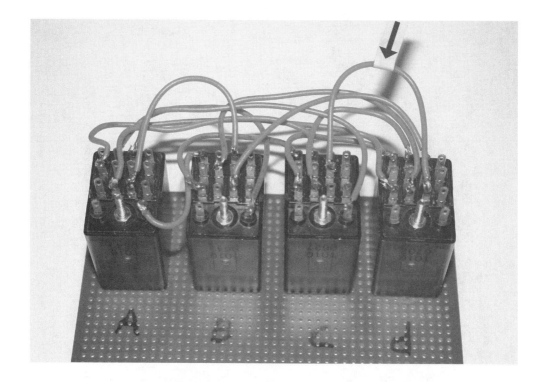

Connect 25 to 26.

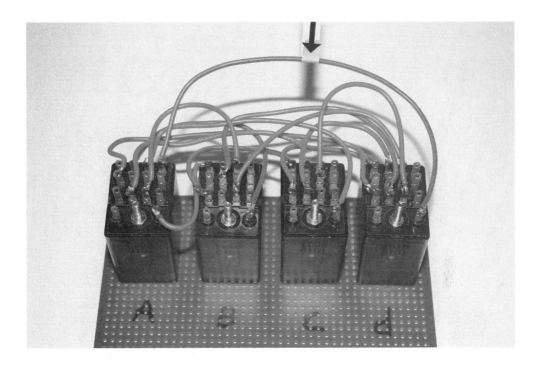

Connect 27 to 28.

Connect 29 to 30.

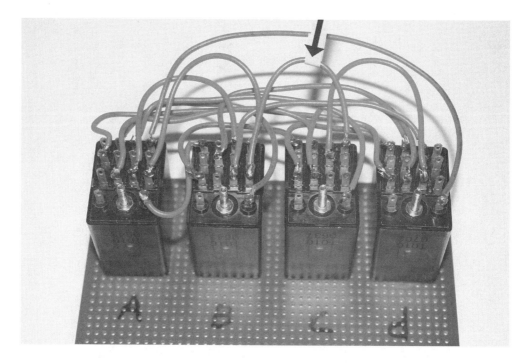

Connect 31 to 32.

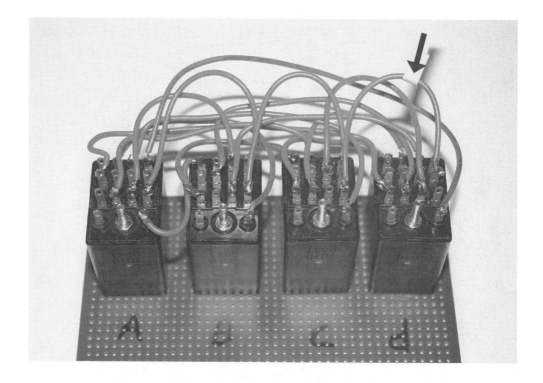

Connect 33 to 34.

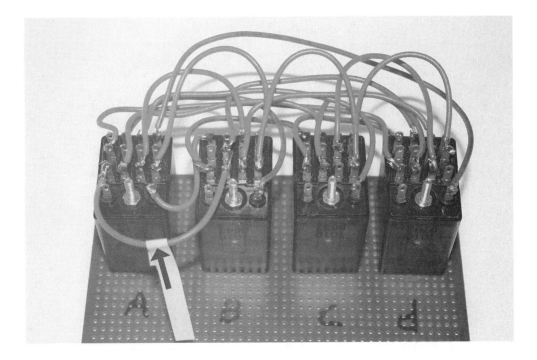

Connect 34 to 35.

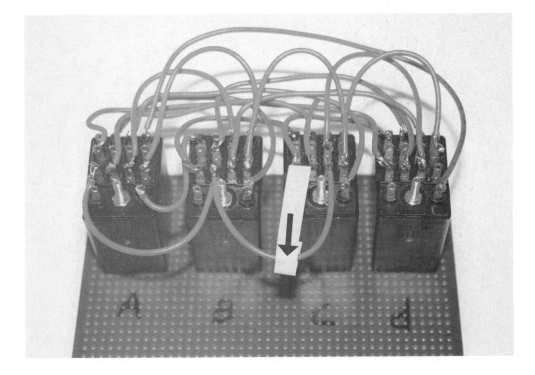

Connect 35 to 36.

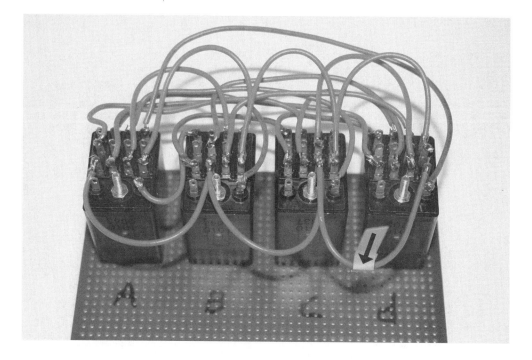

Connect 37 to 38.

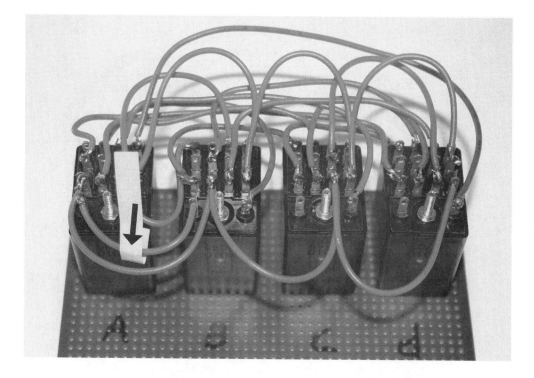

Connect 38 to 39.

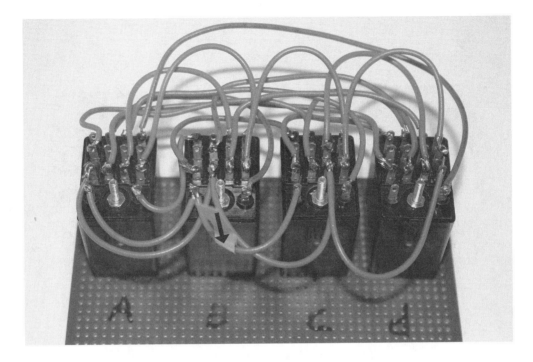

Connect 39 to 40.

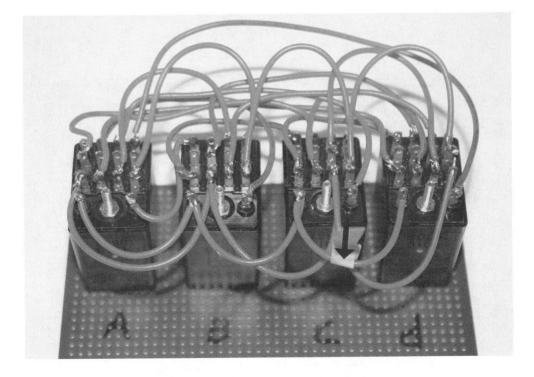

In each of the four switch boxes, drill a ½-inch hole in the top for a switch.

Drill a ¼-inch hole in the side of each of the four switch boxes.

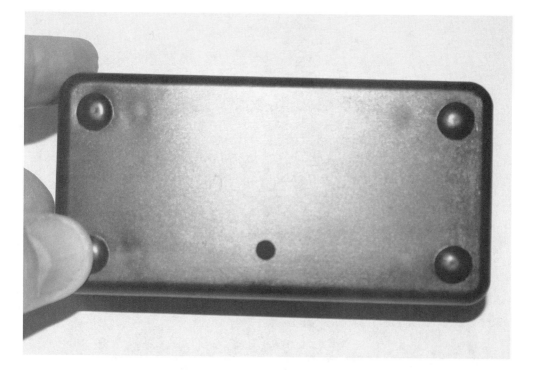

Take a 24-inch length of red wire and a 24-inch length of black wire and twist them together. Strip the ends with a wire stripper. Do this four times.

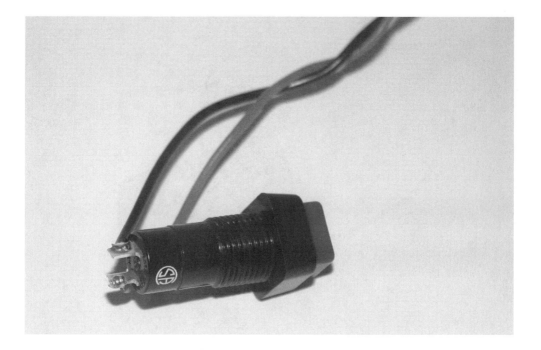

Solder one pair of wires to each of the four push-button switches, red wire to (+) and black wire to (−).

Route the free ends of the wires through the ½-inch top hole in each box, and through the nut that secures the switch.

Route the wires through the ¼-inch hole in the side of each box.

Fasten the switches to each box by tightening the nut. Pull any excess wire through the hole in the side of each box.

Screw the cover onto each box.

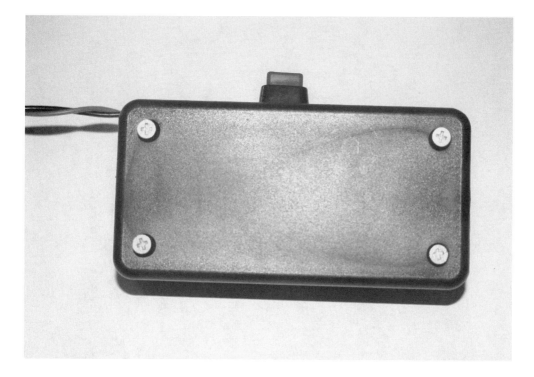

Connect RA to Point 1.

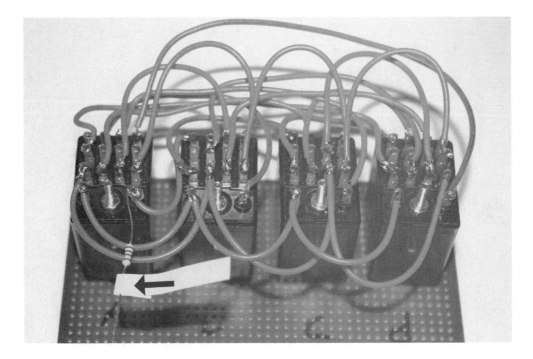

Connect RB to Point 9.

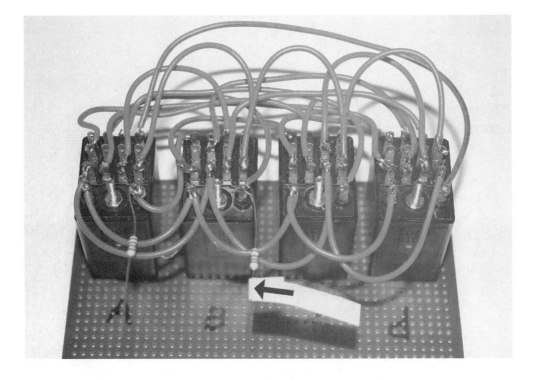

Connect RC to Point 17.

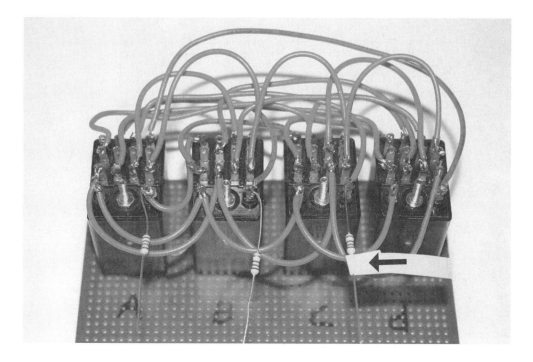

Connect RD to Point 25.

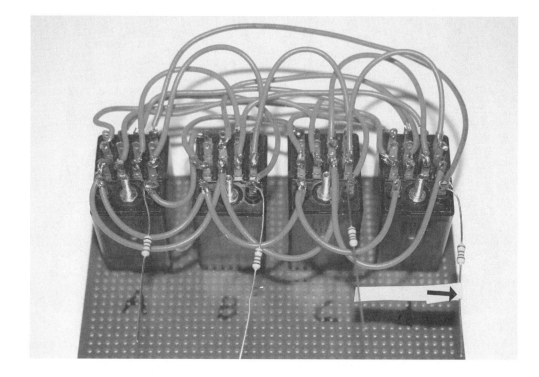

Drill four ¼-inch holes, equally spaced, in the top of the large box. Mount each LED.

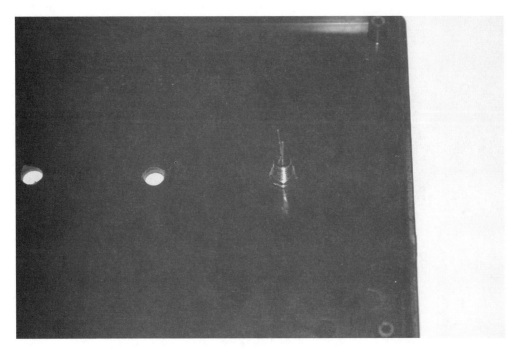

Back

Front

Solder a wire connection from the short lead of each LED to the next LED (Point 41 to 42 to 43 to 44). Solder a 6-inch length of wire from the last LED (44 to 57).

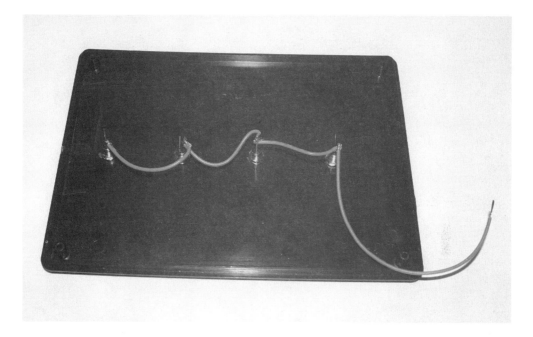

Solder a 6-inch length of wire to the long lead of each LED (Points 45, 46, 47, and 48).

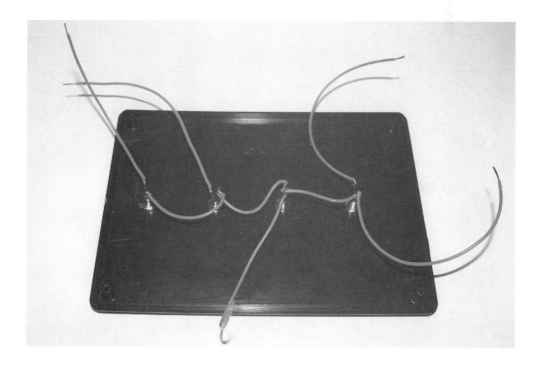

You should have five wires coming from the LED assembly, one to each LED and one that is common (jumps to all the LEDs). Solder each of the five wires to a pin on the connector.

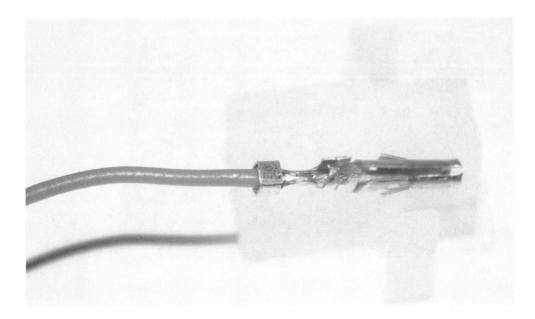

Push the first connector pin into the socket (Point 57 to JM1). See the diagram below for pin positions in the socket.

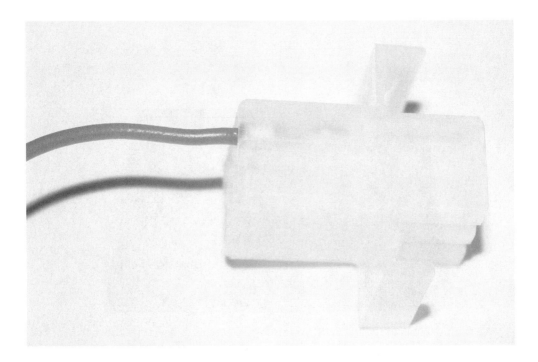

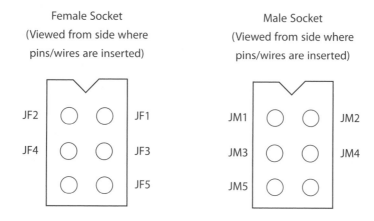

Female Socket
(Viewed from side where
pins/wires are inserted)

JF2   ◯   ◯   JF1

JF4   ◯   ◯   JF3

       ◯   ◯   JF5

Male Socket
(Viewed from side where
pins/wires are inserted)

JM1   ◯   ◯   JM2

JM3   ◯   ◯   JM4

JM5   ◯   ◯

Connect the other pins, JM2, JM3, JM4, and JM5, to the socket.

Glue a barrier strip to the Perfboard.

Attach a piece of Velcro tape to the inside of the large box.

Attach a corresponding piece of Velcro tape to the bottom of the Perfboard.

Drill two holes in one end of the large box, ¼-inch diameter for the reset switch, ⅜-inch diameter for a rubber grommet.

Drill a single ⅜-inch-diameter hole in each of the other sides of the large box (three holes in all).

Push a grommet into each of the three holes. The purpose of the grommets is to protect the wires where they enter the box.

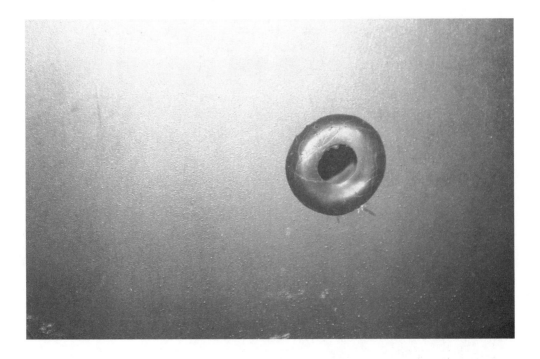

On the end with two holes, mount the switch and a grommet.

Find the end of resistor RA (Point 49).

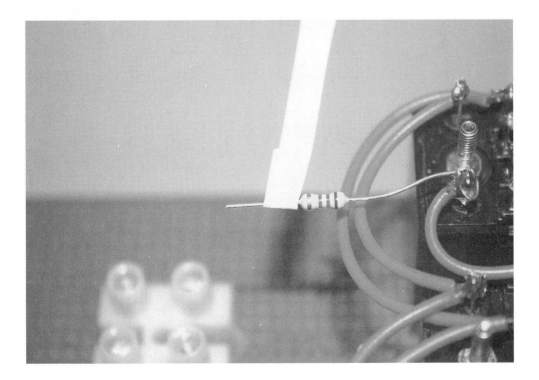

Solder a 4-inch wire to RA (49).

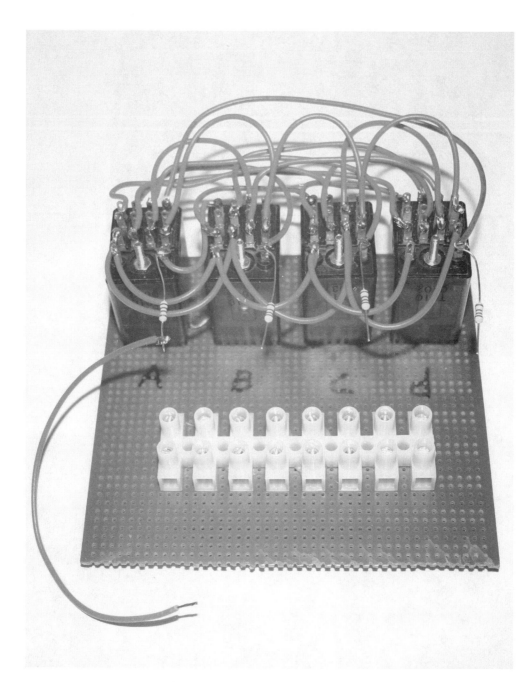

Solder 4-inch wires to RB (50), RC (51), and RD (52).

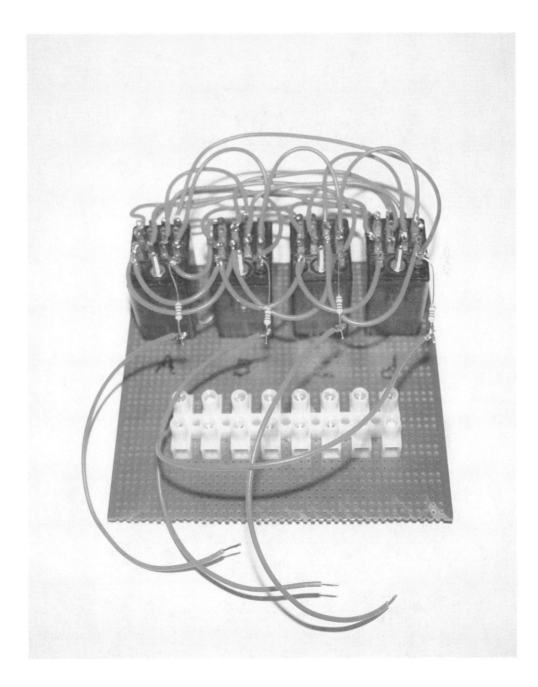

Solder connector pins to RA, RB, RC, and RD.

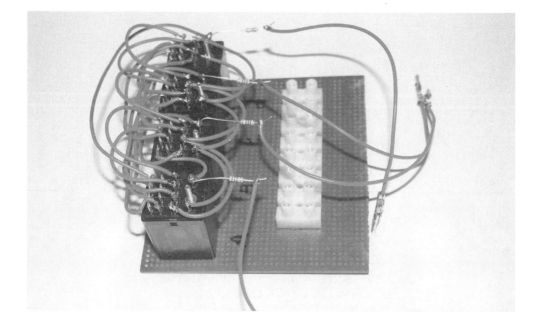

Insert pin (from 49) into connector JF2 (look at the diagram on page 167). Pay close attention here. Because you are looking at the other side of the socket, positions that were on the right side of the LED assembly are now on the left side of the box assembly.

Push JF3 into the connector.

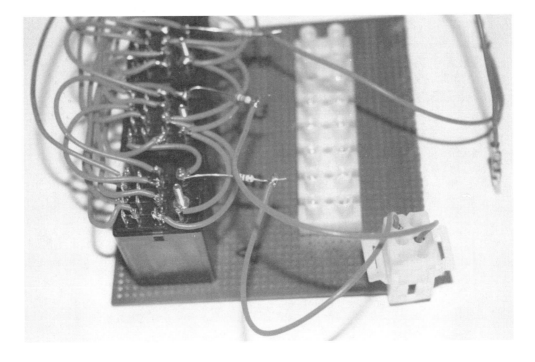

Push JF4 into the connector.

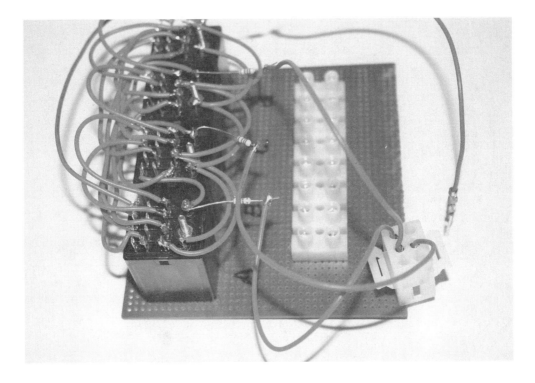

Push JF5 into the connector.

Connect T8 to 36.

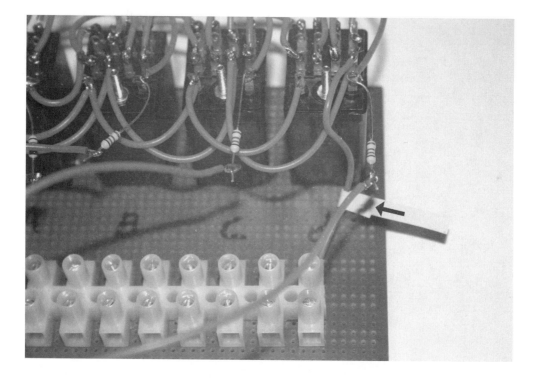

Connect T7 to 40.

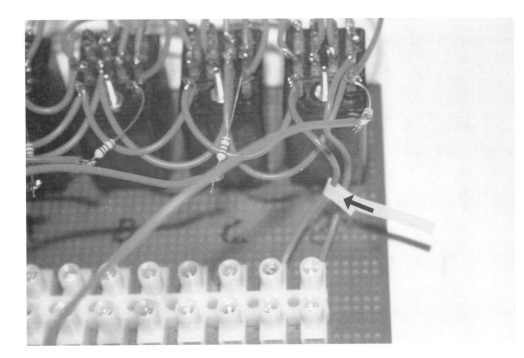

Connect another wire from T7 to JF1.

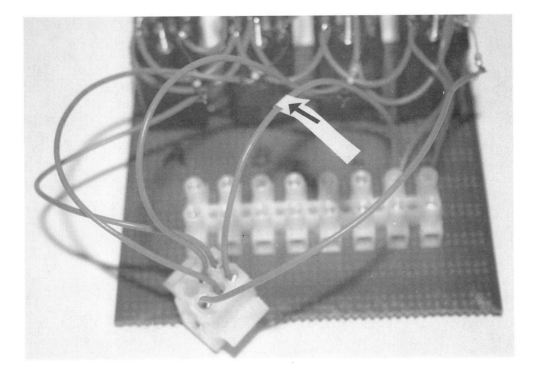

Place the Perfboard into the large box. The Velcro should secure the board to the box.

Bring the power supply leads into the main box (+) to T8 and (−) to T6.

Connect T6 to one side of the reset switch.

Connect the other side of the reset switch to T7.

Connect T8 to T4.

Connect a second wire to T4.

Connect T4 to T3.

Connect T3 to T2.

Connect T2 to T1

Bring Push Button A wires into the large box. Connect the red wire to T1. (It really doesn't matter whether you use red or black, but it's best to stay with me to avoid confusion.)

Connect the black wire from Push Button A to 8.

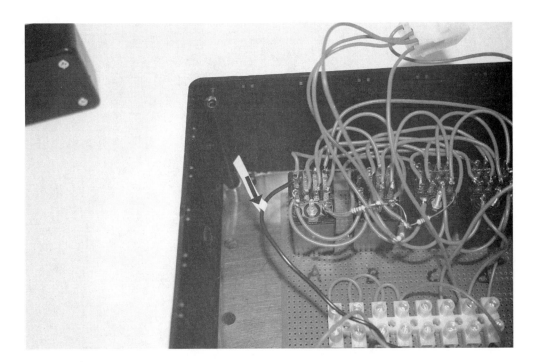

Take the Push Button B wires into the box (through a different hole from the Push Button A wires). Attach the red B wire to T2.

Attach the black B wire to 16.

Bring the Push Button C wires through an unused hole. Attach the red C wire to T3.

Attach the black C wire to 24.

Bring the Push Button D wires through the last empty hole. Attach the red D wire to T4.

Attach the black D wire to 32.

Plug the connectors together and fasten the top with screws.

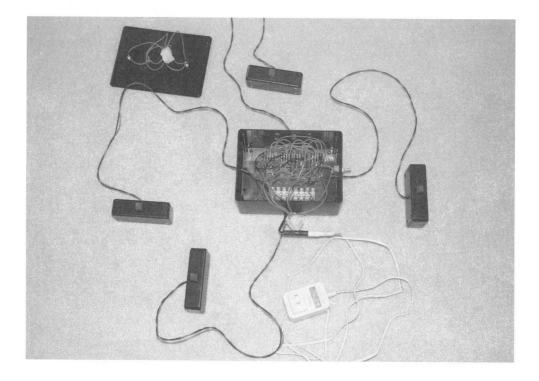

Put labels on the switches and LEDs. Power it up and have fun!

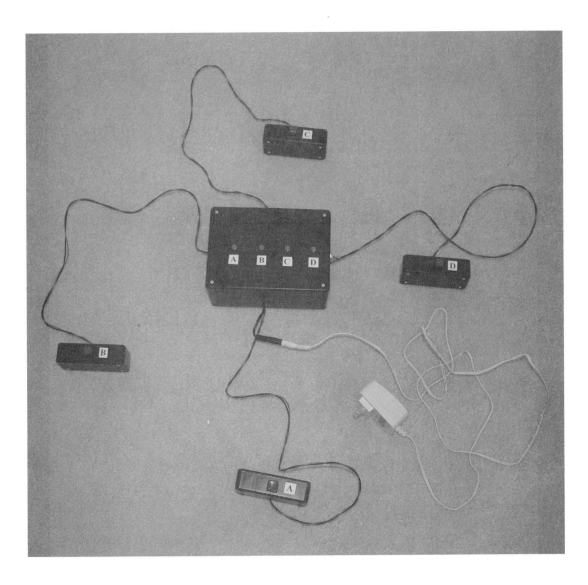

More Projects from Mike Rigsby!

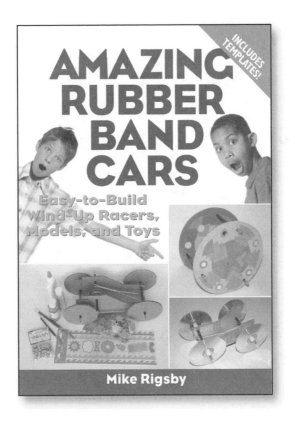

Amazing Rubber Band Cars
Easy-to-Build Wind-Up Racers, Models, and Toys
By Mike Rigsby
978-1-55652-736-4
$12.95 (CAN $13.95)

"Totally irresistible, admirably educational, and excitingly mobile projects."
—*Home Education Magazine*

There's no need for expensive construction kits if you have a handful of rubber bands, a bottle of glue, a pile of recycled cardboard, and a copy of *Amazing Rubber Band Cars*. Mike Rigsby gives you step-by-step instructions to build a variety of racers, including templates. Once readers have built the basic rubber band car, they are given alternatives to modify their designs—replacing the wheels with discarded CDs, installing axle bearings, and improving the rubber band drives to make their cars go even farther. Later designs incorporate steering mechanisms, lifter bars, hinges, and cam shafts—and for those readers completely captivated by cardboard vehicle technology, Rigsby includes plans for a sturdy car large enough to transport a human.

Art of the Catapult
Build Greek Ballistae, Roman Onagers, English Trebuchets,
and More Ancient Artillery
By William Gurstelle
978-1-55652-526-1
$16.95 (CAN $18.95)

"This book is a hoot . . . the modern version of *Fun for Boys* and *Harper's*
Electricity for Boys."   —*Natural History*

Whether playing at defending their own castle or simply chucking pumpkins over
a fence, wannabe marauders and tinkerers will become fast acquainted with
Ludgar, the War Wolf, Ill Neighbor, Cabulus, and the Wild Donkey—ancient
artillery devices known commonly as catapults. Instructions and diagrams illus-
trate how to build seven authentic working model catapults, including an early
Greek ballista, a Roman onager, and the apex of catapult technology, the English
trebuchet. Additional projects include learning how to lash and make rope and
how to construct and use a hand sling and a staff sling.

Backyard Ballistics

Build Potato Cannons, Paper Match Rockets, Cincinnati Fire Kites,
Tennis Ball Mortars, and More Dynamite Devices
By William Gurstelle
978-1-55652-375-5
$16.95 (CAN $18.95)

"If you'd like to launch a potato in a blazing fireball of combusting hairspray, this
is your best source."   —*Time Out New York*

Ordinary folks can construct 13 awesome ballistic devices in their garage or base-
ment workshops using inexpensive household or hardware store materials and this
step-by-step guide. Clear instructions, diagrams, and photographs show how to
build projects ranging from the simple—a match-powered rocket—to the more
complex—a scale-model, table-top catapult—to the offbeat—a tennis ball cannon.
With a strong emphasis on safety, the book also gives tips on troubleshooting,
explains the physics behind the projects, and profiles scientists and extraordinary
experimenters such as Alfred Nobel, Robert Goddard, and Isaac Newton.

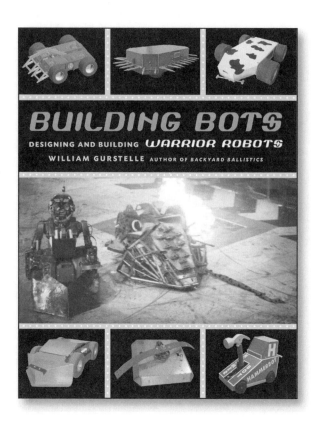

Building Bots
Designing and Building Warrior Robots
By William Gurstelle
1-978-55652-459-2
$19.95 (CAN $29.95)

"Handy how-to guide for turning your garage into a robot war zone."
—*The New Yorker*

This is the definitive guide to designing and building warrior robots like those seen on *BattleBots*, *Robotica*, and *Robot Wars*. It walks robot enthusiasts of all ages step-by-step through the design and building process, enabling them to create any number of customized warrior robots. Chapters include information on designing a robot, choosing materials, radio control systems, electric motors, robot batteries, motor speed controllers, gasoline engines, and drive trains. Clear instructions are accompanied by photos, line drawings, and detailed diagrams throughout. For beginners, there is Machine Shop 101 and robot physics, and, of course, chapters on weaponry that include spinner robots, thwackbots, cutting blade robots, lifters, and chameleon robots.

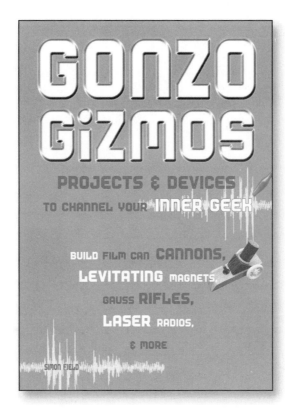

Gonzo Gizmos
Projects & Devices to Channel Your Inner Geek
By Simon Field
978-1-55652-520-9
$16.95 (CAN $18.95)

This book for workbench warriors and grown-up geeks has step-by-step instructions for building more than 30 fascinating devices. Detailed illustrations and diagrams explain how to construct a simple radio with a soldering iron, a few basic circuits, and three shiny pennies. Make a rotary steam engine that requires a candle, a soda can, a length of copper tubing, and just 15 minutes. Roast a hot dog with just a flexible plastic mirror, a wooden box, a little algebra, and a sunny day. Also included are experiments most science teachers never demonstrated, such as magnets that levitate in midair, metals that melt in hot water, a Van de Graaff generator made from a pair of empty soda cans, and lasers that transmit radio signals.